Creating new boundaries in eco-friendly papermaking

TNPL is committed to manufacture eco-friendly paper of World class quality to the complete satisfaction of its customers. The successful manufacture of quality paper from bagasse is an index of the company's technological competence. A strong record in adopting "minimum impact-best process" technology, resource conservation, continuous augmentation of wind energy, responsible waste management and reduced pollution load make the Company one of the most environmentally compliant paper mills in the world.

TNPL's design excellence, quality commitment and system compliance have been validated by the ISO 9001:2000 certification, while its environment management system has been endorsed by the ISO 14001 certification.

TNPL

Paper of the future

Tamil Nadu Newsprint and Papers Ltd.

No, 67, Mount Road, Chennai 600 032, India. Tel: 22354415-17, 22301094-98 Fax: 22350834, 22354614 Email: response@tnpl.co.in

THEMELINE / 608

Environment best practices: industry initiatives

Cleaner is cheaper – Volume 6

Edited by
R K Narang

teri
The Energy and Resources Institute

ISBN 978-81-7993-233-9

Rs 995/-

Published by
TERI Press
The Energy and Resources Institute
Darbari Seth Block
IHC Complex, Lodhi Road
New Delhi – 110 003
India

Tel. 2468 2100 or 4150 4900
Fax 2468 2144 or 2468 2145
India +91 • Delhi (0) 11
E-mail teripress@teri.res.in
Web www.teriin.org

Printed on recycled paper.

Printed at Batra Art Press, New Delhi, India

Contents

SECTION IV *Environment management system*

ANNEXURES

* Please refer to page number 136
** Please refer to page number 17

Foreword

The massive problem of environmental degradation taking place worldwide has added urgency to the need for all stakeholders to pursue their stated goals, whether in government, business or civil society, while at the same time protecting the earth's natural resources. Hence, production and consumption based on clean technologies, processes, and products must be important parts of the agenda for human society if we are not to be overwhelmed by the harmful consequences of continuing environmental degradation. In fact, cleaner indeed is cheaper because the organizations that protect the environment while at the same time producing more also succeed in maximizing their profits. There are several examples of this fact, as indeed one can see from the history of various countries and organizations that met the sudden increase in oil prices, which took place over three decades ago. Those organizations that brought a radical improvement in energy efficiency found that in the process, they were also able to improve their overall productivity and earn much higher profits.

Adding to the strong rationale for producing and consuming in a manner that can be labelled as clean is the knowledge that we now have on the subject of climate change. The IPCC (Intergovernmental Panel on Climate Change) *Fourth Assessment Report*, or AR4, has clearly established the unequivocal reality of climate change, based on extensive observations that show us how human actions are affecting the earth's climate. The world today is in the thick of negotiations, the culmination of which is expected to lead to a global agreement for tackling the challenge of climate change in the Conference of the Parties at Copenhagen. As a result, it is expected that developed countries in particular would have to reduce GHG (greenhouse gas) emissions, and subsequently, developing countries would also have to do so effectively. However, developing countries will have to get active in moving in the same direction very early for two important reasons. First, many of the mitigation measures required for reducing GHG emissions also have large-scale co-benefits at the local level and would promote sustainable development. Hence, while it is not expected that developing countries would reduce their emissions in the short term, they can certainly reduce the intensity of energy use and related GHG emissions significantly. The second reason why early action by the developing countries would be in their interest is that in a rapidly globalizing economy, companies that want a larger share of the market will have to anticipate and deal with the low carbon opportunities that would emerge in the future.

This edition carries a number of case studies, which establish best practices in a number of areas, including water and waste water management, process improvement, product innovation, establishment of effective environment management systems, management of waste in general, as well as conservation of natural resources through the policies and actions of the corporate sector.

India has been uniquely successful in attaining high rates of economic growth in recent decades, but its further success will depend greatly on the ability of corporate organizations to anticipate the future and ensure natural resource use efficiency and environmental protection at unprecedented levels. The case studies presented in this volume clearly represent a strong business case for adopting this approach and establishing that in reality cleaner is cheaper.

R K Pachauri
Director-General, TERI
and
Chairman, IPCC

Acknowledgements

Guidance
R K Pachauri
R K Narang
Annapurna Vancheswaran

Coordination
Core TERI Corporate Awards Team
Pratik Ghosh
Sophia Lonappan
Pooja Kumar
Bhuvan Verma
V B Suneetha

Case study evaluators
Suneel Pandey
Sachin Kumar
Sumit Sharma
Anshuman
Sridhar P Chidambaram
Neha
Sunil Kumar
Shazneen Gazdar
Sonya Fernandes
Ankur Garg
Arupendra Nath Mullick

Core TERI Press Team
Madhu Singh Sirohi
Richa Sharma
Yoofisaca Syngkon Nongpluh

Writing
Richa Sharma

Copy-editing
Richa Sharma
Yoofisaca Syngkon Nongpluh
Gopalakrishnan Srinivasan

Design and production
R Ajith Kumar
Mahfooz Alam
R K Joshi
T Radhakrishnan

Marketing and outreach
Binita Singh
Amit Kumar Ranjan
Prema Mahadevan

Introduction

In the *Approach Paper to the Eleventh Five-year Plan,* the Planning Commission (2006) recognizes environmental concerns as a major challenge for India and suggests that the country's future development strategy has to be sensitive to these growing concerns, and emphasizes that the country should ensure that these threats and trade-offs are appropriately evaluated. Also, the IPCC (Intergovernmental Panel on Climate Change) *Fourth Assessment Report* and the Stern Report have presented incontrovertible evidence of how human activity contributes to global warming and calls for immediate actions to combat this universal challenge. Business and industries would also undoubtedly face several challenges in this regard, all the more in India, with its economy closely linked to its natural resource base and climate-sensitive sectors.

Environmental considerations are likely to form an integral part of commercial normality and, indeed, competitiveness in the future. Definitions of business success are likely to include the assumption of zero-negative impact on the environment at the very least. The political pressure to upgrade environmental standards in India has taken place largely because of the commitments under multilateral environmental agreements and commercial interests, given the new environmental provisions in the multilateral trade regime of the WTO (World Trade Organization). Even before stringent environmental legislation comes into force, businesses should recognize that it is not only ethical to be environment friendly, but with the growth in consumer awareness in the environmental area, it will also be good business. It is most obvious that the prerequisites for businesses to survive through the middle years of this century necessitate a critical review of their direct and indirect use of energy and ways of reducing it drastically. If they do not do this, they will be passive victims of price increases, quota restrictions, and implicit or explicit carbon taxes and other measures to contain emission growth.

The World Bank (2007) in its *Country Environment Analysis for India* refers to the history of environmental management worldwide, which points out that effective environmental enforcement requires informed consensus on environmental management objectives and policies that are based on a good understanding of the shared roles and responsibilities of all players, including the regulator, regulated community (developers and polluters), and affected community (general public). Companies are beginning to realize that environmental issues need to be addressed for a number of reasons, including consumer pressure, potential cost savings, legislation, and ethics.

However, companies are faced with the challenge of integrating environmental considerations into their production and marketing plans. Research on corporate strategy and the environment suggests that there are three key options available to organizations–improving market communication, improving manufacturing processes, and carrying out research and development. The traditional value chain to the environment may also be adapted to identify the various ways for the development of competitive environment strategies. A general policy has to be developed by the organization before any environmental management activity. An environmental review provides a snapshot of the environmental performance of the organization and enables the development of realistic policies and recommendations. The other helpful factors are training, allocation of responsibility, setting and attainment of targets, operational control and audit, environmental reporting and environmental statements, standardization, structure of the management system, and so on.

The Indian industry is now taking a more proactive stance and is recognizing that sound environmental management on a voluntary basis can enhance corporate image, increase profits and competitiveness, reduce costs, and obviate the need for further legislative measures by regulatory authorities. Evidence of this is seen in the move towards 'green products', with the increasing use of 'life cycle analysis'; that is, looking at the environmental impact of a product 'from cradle to grave'. The industry has also developed a number of environmental management tools, such as environmental auditing and environmental management systems. Research findings on the relationship between financial and environmental performance reveal that the level and quality of environmental reporting by a firm are correlated with its environmental performance, and better reporting of data is associated with better management and transparency, which makes a firm more attractive to investors.

Though the implementation of structured environmental management in companies remains voluntary, companies are carefully assessing not only the potential financial benefits (identification of savings, improved production efficiency, new market potential, and so on) that may arise from such activities, but also the risks of not addressing organizational as well as technical solutions to environmental problems (including accidents, inability to obtain bank credit and private investment, loss of markets and customers).

The broad objective of this compendium is to bring to the public domain some of the best practices in the Indian industry, so as to inspire individual organizations to learn from the experiences of other organizations and enable replication of such initiatives on a larger scale. The case studies highlight the innovative strategies used by corporations to address the environmental challenges and how these have helped the companies financially while reaffirming the fact that it pays to be environment friendly. Some of these solutions would also provide opportunities for the expansion of the role of business.

The compendium contains case studies from some of the leaders of the Indian industry, both in the public and private sectors. The case studies have been classified into the following four broad categories.

1 Water and waste water management
2 Waste management and resource conservation
3 Process improvement and product innovation
4 Environment management system

Many of the case studies, however, actually have cross-cutting themes and objectives and incorporate more than one of the above categories, thereby reflecting the integrative character of their efforts. Certain effective environmental management systems and sustainable resource utilization approaches are multidisciplinary tasks.

SECTION

I

Water and waste water management

The water sector in India faces major challenges with declining per capita availability. Water demand from agriculture, domestic, and industrial sectors is rising due to growing population and developmental activities. Water pollution further complicates the situation. Further, there is a potential threat of climate change impacts on the availability of water. In future, the direct and indirect costs of water would increase significantly, and, at the same time, quantity and quality of water would be far more challenging for industrialization.

As the challenges magnify, water management will take the centrestage in the planning process followed by the coporates, such that it remains in a coherent harmony with the environment. The contemporary approach of typical 'end-of-pipe' treatment of industrial waste water needs to shift towards decentralized, process integrated water management, with efforts concentrated towards 'zero discharge' or 'positive water balance', thus reducing the demand for fresh water. There is a need for adopting water management practices that integrate the use of water-efficient technologies and treatment systems and which include mandatory water conservation efforts such as rainwater harvesting, groundwater recharge, waste water reuse and recycle, and water auditing. The basic necessity is to save and preserve precious water for sustainable use amongst different users.

The Indian industries are moving towards excellence in their water management practices gradually and continuously. There is a growing realization amongst the Indian corporates for efficient water management that moves beyond regulatory compliance and improves their water conservation efforts, demonstrating a sense of green business.

The current section highlights some of the case studies that demonstrate appreciable efforts to promote rainwater harvesting, natural recharge of groundwater, zero discharge of effluent, waste water reuse and management, and improvement in water consumption.

CASE STUDY 1

Hindustan Aeronautics Ltd

Figure 1 Entrance of the Engine Division of the HAL (Hindustan Aeronautics Ltd) in Bangalore

About the company

The Engine Division of the HAL (Hindustan Aeronautics Ltd) was set up in 1957 in Bangalore to manufacture Orpheus turbo jet engines under license from Rolls Royce (Figure 1). In 1959, the Division signed another license agreement with Rolls Royce to manufacture Dart engines to power HS-748 passenger aircraft and overhaul Avon engines for Canberra and Hunter aircraft. Since then, the Division has spread its wings and now manufactures Artouste engines for Chetak/Cheetah helicopters, Adour engines for Jaguar aircraft, and Garrett engines for Dornier aircraft. The Division also undertakes repairing and overhauling works for various engines operated by Indian Air Force, Indian Navy, Indian Army, Coast Guard, Border Security Force, corporate sector, state government, and others. The Division is equipped with the state-of-the-art technologies to manufacture, repair, and overhaul aircraft engines. The Engine and Test Bed R&D (research and development) Centre, a part of the Engine Division, specializes in the development of small gas turbines and engine test beds.

Environmental profile

The company has an Environment Policy at the corporate level. It has a combined Safety, Health and Environment Department at the corporate level. The Department strives to integrate environment-friendly practices into the company's work culture. Some of its functions are as follows.

- Maintaining sewage treatment plant, effluent treatment plant, scrubbers, bag filters, cyclone separators, and other pollution control equipment.
- Ensuring compliance with all statutory requirements for improving the environment.
- Implementing cleaner production and pollution prevention techniques.
- Managing hazardous waste.
- Addressing all environment issues.
- Providing personal protective equipment to all personnel.
- Generating awareness and organizing training programme for the employees on safety, health, and environment.

The company is ISO 14001: 2004 certified and has an EMS (Environment Management System) in place at the corporate level.

Air pollution

The major air pollutants emitted from the Engine Division are SPM (suspended particulate matter), RSPM (respirable suspended particulate matter), SO_2 (sulphur dioxide), NO_x (oxides of nitrogen), acid mist, oil and coolant fumes, and odour.

Air pollution control measures

SPM and RSPM emissions

SPM and RSPM are released during blasting and grinding operations, and while polishing machines, and spray painting booths. To control these emissions, dust collectors, cyclone separators, and water curtains have been installed.

SO_2, NO_x, and acid mist emissions

The process of electroplating and chemical cleaning releases oxides of sulphur and nitrogen, and acid mist. Wet scrubbers have been installed to control these emissions.

Oil and coolant fumes

Oil and coolant fumes are generated from machines that use oil as lubricants and coolants to prevent overheating. Such machines are equipped with electrostatic fume killers, which trap the fumes, preventing their release into the atmosphere.

Odour

Rig testing areas used to generate kerosene odour. To contain this problem, these areas are provided with an air ventilation system comprising an exhaust system and filters.

Noise pollution

Test beds and blasting machines generate high level of noise. Personnel working in such high noise areas are provided with ear plugs and ear defenders. Acoustic enclosures are also provided in some places.

Water pollution

Domestic waste water and its control measures

Domestic waste water generated from toilets, canteens, and kitchen contains pH, BOD (biochemical oxygen demand), TDS (total dissolved solids), TSS (total suspended solids), oil and grease, chlorides, sulphates, total Kjeldhal nitrogen, ammonia, nitrogen, and so on.

This waste water is treated in a sewage treatment plant, which has various treatment units, including a unit for activated sludge process and a tertiary treatment unit comprising pressure sand filter and activated carbon filter. The treated water is used for horticultural purposes.

Facility effluent and its control measures

The effluent generated from plating shops and chemical cleaning shops contains pH, BOD, COD, TDS, TSS, chlorides, sulphates, and heavy metals like cadmium, chromium, cyanide, nickel, and so on.

The effluent stream is segregated into two streams at the source: chromium effluent and cyanide effluent. The chromium effluent is treated using chromium reduction method, followed by precipitation method using sodium meta bisulphate and NaOH (sodium hydroxide). The cyanide stream is treated using sodium meta bisulphate and NaOH in an oxidation method. Tertiary treatment of these effluents is carried out using carbon and sand filters. The treated water is mixed with treated domestic waste water, which is used for horticultural purposes.

Hazardous waste

The hazardous waste generated from the facility and its disposal mechanism are listed below.

- Used oil is generated, which is auctioned to the CPCB (Central Pollution Control Board)-authorized recyclers.
- Paint residue, sludge from the effluent treatment plant, and spent hardening salt recovered from the cyanide treatment process are collected in PVC (polyvinyl chloride) drums, covered with polythene covers, and stored in RCC (reinforced cement concrete) rooms. As per the KSPCB (Karnataka State Pollution Control Board) directions, preliminary action has been taken for the disposal of hazardous sludge to CHW-TSDF (common hazardous waste-treatment storage disposal facility).

Other waste

Solid waste from canteens and ferrous and non-ferrous scraps from machining area are other types of waste generated from the facility. The canteen waste is sent to pig yards, where it forms the feed for pigs. Ferrous and non-ferrous scraps are stored in a designated place. The company is in the process of procuring a scrap baling machine for making bales, which will be e-auctioned by a Government of India enterprise.

Case study

Installation of a rainwater harvesting and artificial recharge system at the Engine Division

About rainwater harvesting and groundwater recharge

Increasing population and agricultural demands in India are exerting immense pressure on existing water resources, which are depleting fast. There is a high percentage of population that does not have access to freshwater resources. So, there is an urgent need to conserve and augment water. The solution to this problem lies with nature. Nature provides mankind with abundant fresh water in the form of rain. There is a need to collect this rainwater and use it for various purposes. So, rainwater harvesting is essentially the collection, filtration, and storage of rainwater. The surplus water after storage can also be used for recharging groundwater through artificial recharge techniques.

Scope of rainwater harvesting in the Engine Division

The Engine Division has developed a green belt in an area of 2.4×10^4 m^2, which requires huge quantity of water. In a bid to save fresh water that was being largely used to water and nurture the green belt, the Division installed a rainwater harvesting system.

After calculating the amount of rainfall received by Bangalore, rooftop area, and area that could be used for rainwater harvesting, it was concluded that 5 644 000 litres of rainwater could be harvested per annum. Keeping the economics in mind, a tank of 77 000-litre capacity was constructed for storing rainwater.

A sketch of the rainwater harvesting system at the Engine Division

The rainwater harvesting system comprises a filter tank and a storage tank.

Filter tank

Rainwater from the rooftop of the machine shop area is made to flow through a PVC pipe to the MS (mild steel) sand filter tank, where it is filtered (Figures 2 and 3). This filter tank is mounted on I section, which is fixed on a 200-mm-thick RCC bed. The size of the tank is

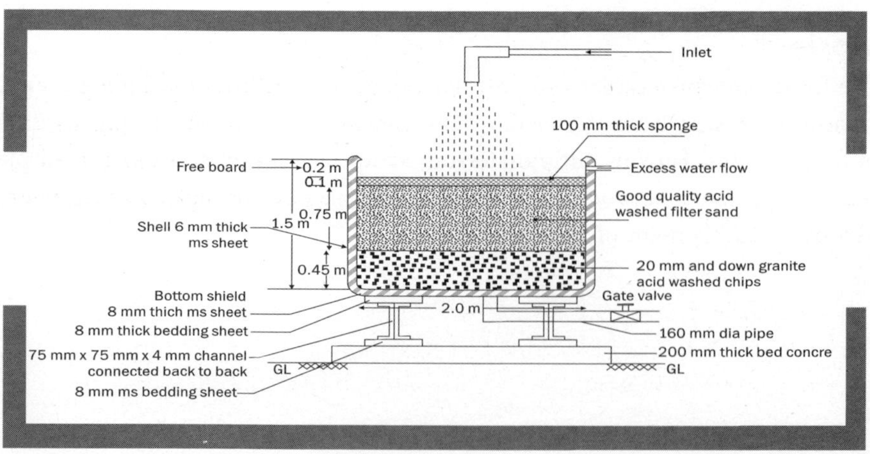

Figure 2
Cross-sectional view of an MS filter tank

Figure 3
View of the filter tank with inlet and outlet arrangements

2 m × 2 m × 1.5 m, including a free board of 0.2 m. The filter tank has two media: 750-mm deep filter sand and 450-mm deep down granite acid-washed chips. The side walls and bottom of the tank are fabricated with 6- and 8-mm-thick MS sheets, respectively, which provide adequate strength to the tank. The top of the filter media is covered with a 100-mm thick sponge, which, in turn, is covered with an MS mesh.

Storage tank

The filtered water flows to the storage tank through two 160-mm diameter PVC pipes equipped with gate valves. The storage tank has a dimension of 11 m × 5 m × 1.4 m, including a free board of 0.6 m. It comprises a manhole of size 600 mm × 600 mm and a pumping system. An air vent pipe with weather cowl is a part of the storage tank.

Sprinkler system

The stored water is used for watering the green belt with a sprinkler system, every 30 minute interval.

Infiltration well

The excess water from the storage tank flows into an infiltration well that has a dimension of 3 m × 4 m × 3 m. The infiltration well is constructed close to the overflow point and filled with graded gravel for artificial recharging the groundwater.

Figure 4 gives a schematic of the rainwater harvesting and artificial recharge system. Figure 5 shows a photographic view of the rainwater harvesting and artificial recharge system.

Cost–benefit analysis

Total investment made in installing the rainwater harvesting system: Rs 981 359
Total fresh water saved per annum: 5 644 000 litres
Total savings per annum: 5 644 000 litres × Rs 0.54 = Rs 3 047 760
Payback period: four months

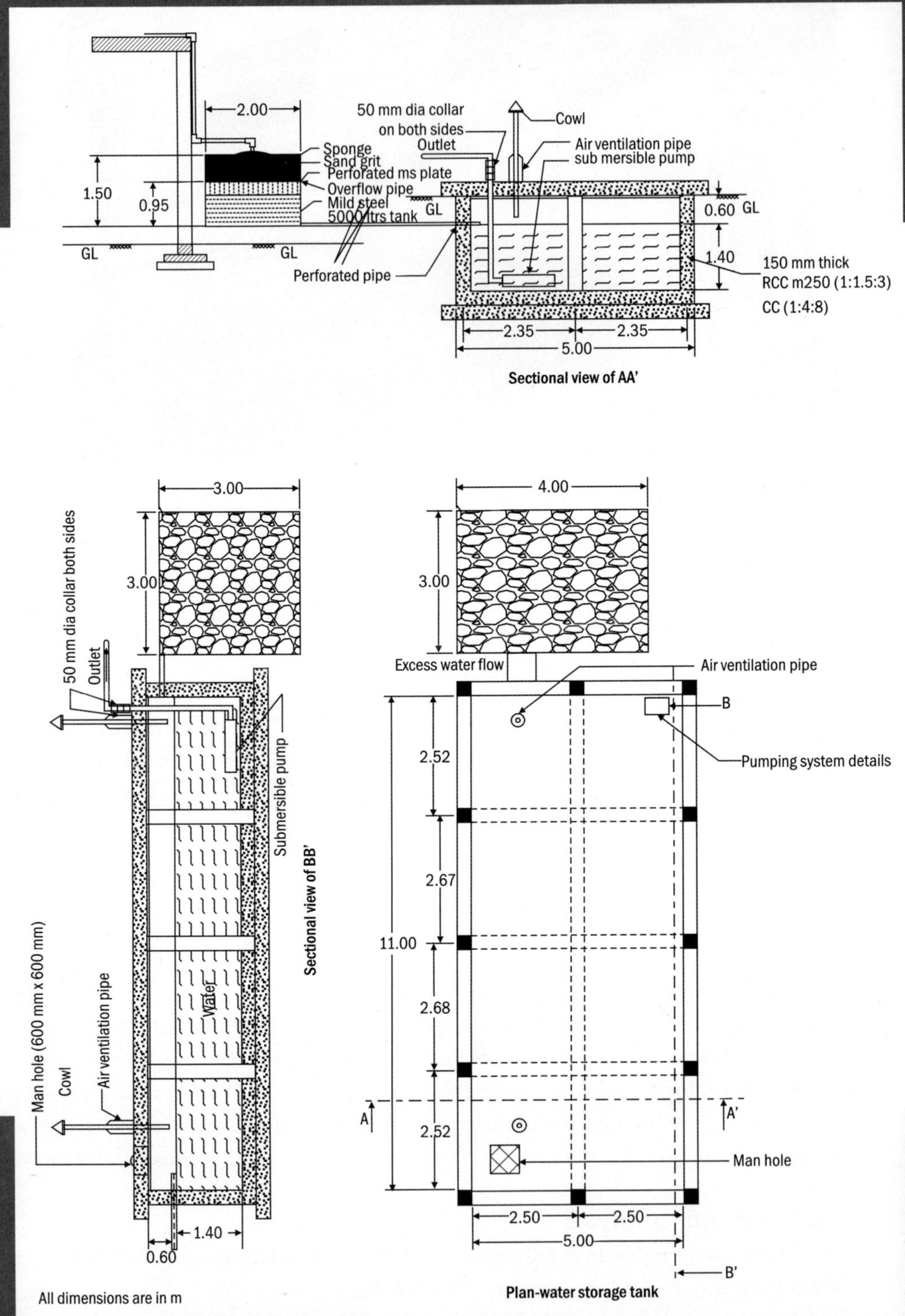

Figure 4 Rainwater harvesting and artificial recharging system

Figure 5 View of rainwater harvesting and artificial recharge system

Benefits

- *Conservation of resources* Since natural water is available free of cost for watering purpose, about 5644000 litres of water is saved per annum.
- *Economics* Saving of 5644000 litres of water per annum results in the saving of Rs 3047760 per annum.
- *Groundwater recharge* Groundwater is recharged through artificial recharge techniques.

CASE STUDY AT A GLANCE

- The Engine Division of the HAL (Hindustan Aeronautics Ltd) was set up in 1957 in Bangalore.
- The Division is equipped with the state-of-the-art technologies to manufacture, repair, and overhaul aircraft engines.
- The Engine Division has developed a green belt in an area of 2.4×10^4 m^2, which requires huge quantity of water.
- In a bid to save fresh water that was being largely used to water and nurture the green belt, the Division installed a rainwater harvesting and artificial recharge system.
- The rainwater harvesting system at the Engine Division comprises a filter tank and a storage tank.
- The excess water from the storage tank flows into an infiltration well, which is used for recharging the groundwater.
- With the installation of the rainwater harvesting system, the company saves 5644000 litres of water per annum, resulting in a saving of Rs 3047760 per annum.
- The groundwater in and around the Division gets recharged.

CASE STUDY 2 Numaligarh Refinery Ltd

About the company

Nestling in the sylvan environs of the Brahmaputra valley in the district of Golaghat (Assam), where the beautiful rendezvous of water and land throws up myriad colours, the NRL (Numaligarh Refinery Ltd) was set up in August 1985, and it has been conceived as a vehicle for speedy industrial and economic development of the region.

The 3-MTPA (million tonnes per annum) NRL was dedicated to the nation by the erstwhile Hon'ble Prime Minister Shri A B Vajpayee in July 1999. The NRL has been able to display creditable performance since the commencement of commercial production in October 2000. With its concern, commitment, and contribution to socio-economic development of the state, combined with a track record of continuous growth, the NRL has been conferred the status of Mini Ratna PSU (public sector unit).

Its product range includes LPG (liquefied petroleum gas), naphtha, motor spirit, aviation turbine fuel, superior kerosene oil, high speed diesel, raw petroleum coke, calcined petroleum coke, and sulphur. The present authorized capital of the company is Rs 10 billion and paid up capital is Rs 7.36 billion.

Environmental profile

The NRL has been pursuing a focused programme towards environment management. It has formulated a comprehensive policy that underscores towards its commitment for the protection, preservation, and development of the environment.

The company strictly adheres to the requirements of the EMS (Environment Management System), conforming to the internationally recognized standard ISO 14001: 2004. In its short period of operation, the NRL has developed and implemented a sound and world class EMS, for which it has been recertified by DNV (Det Norske Veritas) under ISO 14001: 2004 (recertified over 14001: 1996), along with ISO 9001: 2000 and OHSAS 18001: 1999 for Quality Management System and Occupational Health and Safety Management System. The company was awarded ISRS (International Safety Rating System) Level-8 by M/s DNV in the first attempt itself and Level-9 within a span of one year.

Air pollution

The main air pollutants emitted from the facility are SPM (suspended particulate matter), RSPM (respirable suspended particulate matter), SO_2 (sulphur dioxide), NO_x (oxides of nitrogen), and fugitive dust and gases.

Air pollution control measures

- The major fuels used in the Refinery are internally produced sweet fuel gas and naphtha, which have very low sulphur content, resulting in low SO_2 emissions to the atmosphere. It has been proposed to replace naphtha with natural gas having very less sulphur content.
- The NRL is the first Refinery in the North East region to install an SRU (sulphur recovery unit) in spite of processing low sulphur Assam crude (having about 0.22% to 0.26% sulphur), along with sour water stripping unit and sour gas sweetening unit (Figure 1).

Figure 1
Sulphur recovery unit

- Low-NO_x burners have been installed in all furnaces to minimize pollution because of NO_x.
- A first-of-its-kind non-illuminating ground flare system has been installed in the refining industry.
- All furnace stacks have facilities for continuous monitoring of SO_2 and NO_x.
- Regular monitoring of ambient air quality is carried out at five designated locations in and around the Refinery. The parameters monitored are SO_2, NO_x, SPM, CO (carbon monoxide), and HC (hydrocarbon).
- Additionally, 8-hourly TLU (threshold limit value) for SO_2, H_2S (hydrogen sulphide), CO, and Cl_2 is also monitored regularly as part of the work environment monitoring activities.
- To keep the ground-level concentration of pollutants at minimum, the height of stacks in all furnaces has been kept at 60 metres or more.
- Passing valves are monitored by acoustic leak detector, and fugitive emissions are monitored by gas measuring instrument.
- CAT degas process has been implemented in the SRU to remove the entrapped H_2S in liquid sulphur, without exposing to the atmosphere.

- A wind monitoring system has been installed for generating information on wind movements.
- An ultrasonic mass flow meter to measure flare gas has been installed.

Noise pollution control measures

As green belt is an effective barrier for noise, a meticulously planned 100-metre wide green belt around the Refinery and 25-metre wide belt around the marketing terminal have been developed, covering a total area of 60 hectares of land. Further, the Refinery uses low noise equipment and monitors noise on a regular basis.

Water pollution

The main water pollutants discharged are BOD (biochemical oxygen demand), COD (chemical oxygen demand), pH, oil and grease, TDS (total dissolved solids), and heavy metals.

Water pollution control measures

- Adequate segregation, collection, and centralized treatment facilities for waste water have been provided to meet the standards laid down under MINAS (Minimal National Standards) and by the pollution control board.
- A modern ETP (effluent treatment plant) with physical, chemical, biological, and tertiary treatment facilities, using latest technology having three-stage oil recovery system, has been installed (Figure 2).
- In-build measures have been taken for the minimization of waste water generation by reusing the part of the same within the process units. To conserve fresh water and to reduce the quantity of treated effluent, the treated effluent is partly reused in the cooling tower and for gardening purpose.

Figure 2
Effluent treatment plant

Hazardous waste

The main categories of hazardous waste generated from the Refinery are slop oil (waste oil) from all process units and marketing terminal, oily sludge generated from the ETP, and spent catalyst.

Hazardous waste management

- *Recycling of slop oil* Slop oil generated from the process units is drained to the closed blowdown system, from where the oil is directly transferred to wet slop tanks in the off site area. The water accumulated after necessary settlement is drained to the ETP for treatment. Then the dry slop oil is transferred to the dry slop tank for reprocessing.
- *Recycling of waste oil from OWS (oily water sewer) system* The process waste, tank bottom drains, desalter effluents, and effluents from the marketing terminal are routed to the ETP through the OWS system with three-stage oil recovery. The slop oil is taken either to delayed coker unit for reprocessing or transferred to crude tanks for further processing.
- *Recovery of oil from sludge* Oily sludge generated in the different stages of the ETP is collected in a sludge thickener through a sludge collection system and from the thickener. Floating oil with water is recycled back to the inlet receiving sump for further processing and recovering of oil. The oil-free cake from the centrifuge is disposed of in secure landfill.
- *Spent catalyst* Spent catalysts are disposed of through the CPCB (Central Pollution Control Board)-approved recyclers.

Other industrial waste

Various types of industrial waste are generated, as mentioned below.

- Metallic waste, including all types of metallic scrap, both ferrous and non-ferrous, like steel scraps, aluminium, brass, metallic gaskets, and so on.
- Incinerable waste, including waste paper cotton, cloth, ropes, hand gloves, and so on.
- Non-incinerable waste, including oil-soaked insulation, glass wool, and glass waste like fused bulbs, batteries, and so on.

Case study

Re-routing of the sewage treatment plant effluent to the aeration tank in the effluent treatment plant of the Refinery

The STP (sewage treatment plant) located at the NRL township treats domestic sewage originating from the township (Figure 3). The domestic sewage contains large amount of organic matter and suspended solids of different origin. It is made sure that the final effluent in the STP meets the requisite standards before being discharged to inland surface water. The plant is designed for treating 800 m^3/day of domestic sewage.

Raw sewage flow

Average flow: 800 m^3/day, or 33.34 m^3/h

Peak flow: 1600 m^3/day, or 66.67 m^3/h

Figure 3
Sewage treatment plant

Raw sewage characteristics

BOD_5 at 20 °C: 250 mg/l
Suspended solids: 480 mg/l

Treated sewage characteristics

BOD5 at 20 °C: < 10–15 mg/l
Suspended solids: < 20 mg/l

Earlier, the Refinery effluent was discharged into the River Dhansiri after making sure that it met MINAS. Before disposal, it used to get mixed with the effluent from the STP at township. As a water conservation measure, treated effluent was proposed to be reused instead of disposing it of to the river. Some modifications have been made inside the Refinery premises to reuse the effluent as fire water make up, filter blowdown in the cooling tower, and for watering saplings in the green belt zones. Zero discharge, or 100% recycling of treated effluent from the Refinery, has been achieved.

Adopting a further stringent measure towards zero discharge, the township STP effluent was diverted to the ETP, using the redundant pipeline already existing. The STP effluent is fed in the aeration tank in the ETP with a 6-inch pipeline (Figure 4).

The RCC constructed aeration tank is equipped with six surface aerators, which provide both aeration and mixing in the tank. Aerobic conditions are maintained in the tank by the entrainment of the atmospheric air, using surface aerators. The detention period in the tank has been envisaged to be 16 hours. The aeration system is activated sludge process designed under extended aeration mode in which micro-organisms are in endogenous phase. This results in the rapid growth of micro-organisms and, consequently, ensures good quality of treated effluent and settleability of bio-sludge. The waste water is treated with micro-organisms suspended uniformly within the reaction tank to remove biodegradable organic material contributing to BOD/COD. The micro-organisms are maintained in the reaction tank by recycling activated sludge, which is formed by the micro-organisms in the process.

Figure 4
Aeration tank in the effluent treatment plant

Bacteria normally dominate the activated sludge community. The major part is aggregated in sludge flocs, but some are freely dispersed in the liquor. Apart from bacteria, other types of organisms in the activated sludge process are protozoans, ciliates, rotifers, and so on.

The effluent flowing from the aeration tank is routed under gravity to the final clarifier provided with a sludge scrapper, which revolves slowly to scrap the bio-solids settled at the bottom (Figure 5). The clarified effluent stream after biological treatment from the final clarifier is received in the filter feed tank under gravity, from where it is pumped to sand filters.

Activated carbon filtration is an effective means for treating the effluent of undesired colour and odour as well as residual COD. Treated and clear effluent is then routed to guard

Figure 5
Clarifier

ponds for storage. From the guard pond, the effluent is taken for reuse as fire water make-up in the cooling tower filter backwashing and for watering the green belt.

Cost–benefit analysis

The cost incurred in the implementation of the project was Rs 60 000.

Benefits

- By implementing the project, 100% reuse of treated effluent has been achieved.
- The project aids in conserving natural resources by reducing raw water consumption.
- The project also prevents pollution and contamination of waterbody, as there is no discharge of effluents into the waterbody.

Case study

Installation of telescopic insulation for catalyst tubes at the reformer top in the hydrogen unit to cover the bare hot tube length

Process description

There is a 38 000-TPA (tonnes per annum) hydrogen plant operating at the Numaligarh Refinery. The plant was commissioned in February 2000. The plant has a steam reformer, which is a side-fired furnace having 150 catalyst-filled tubes.

There was a significant heat loss from the un-insulated portion of catalyst-filled tubes at the top of the reformer. The surface temperature of these tubes was maintained up to 550 °C. During the normal condition, the operating temperature of the steam reformer is 950 °C, which declines during start-up and shut-down periods. This operating temperature variation in the reformer causes the expansion and contraction of catalyst tubes. This makes it very difficult to provide conventional insulation to cover the exposed portion of catalyst tubes, especially when it is expanded.

Telescopic seals as insulating material

To address this problem, the company procured telescopic seals made up of two overlapping stainless steel shells designed to compensate 310 mm of vertical movement in hot and cold conditions. The shells are filled with a 19 mm layer of ceramic fibre blanket, which has high insulation properties. This telescopic insulation resulted in the saving of considerable amount of heat and fuel. Figure 6 shows the tubes covered with the insulation material.

Cost–benefit analysis

Fuel saved per annum

Naphtha: 400 kl

Total fuel saved in terms of heat content: 3135 Mkcal

Total saving per annum: Rs 17.6 million

Investment incurred in implementing this project: Rs 2.1 million

Payback period: 8.3 years

The arrangement is shown below (Not to the scale):

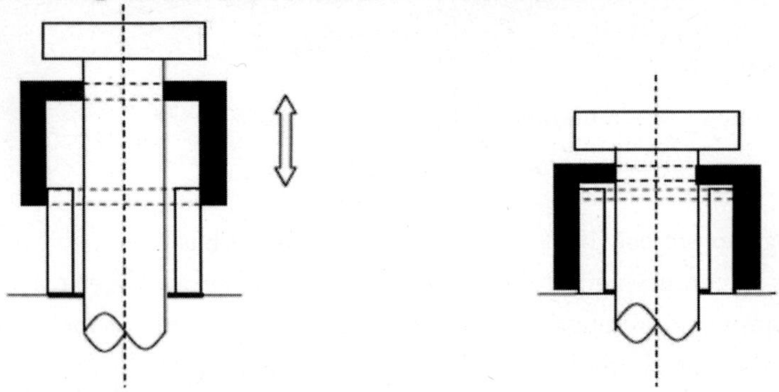

Hot Position (with telescopic seal) Cold Position (with telescopic seal)

Figure 6
Tubes covered with insulation material

Benefits

- *Reduction in heat loss* After installing telescopic seals, heat loss was reduced considerably. There is also a reduction in uncontrolled air volume penetrating the radiant boxes of reformer furnace. As the temperature of the reformer furnace and other conditions are maintained, the excess air penetration would interfere with the conditions and hence the efficiency of the system.
  Surface temperature of the bare tube (before providing insulation): 500 °C
  Surface temperature after providing insulation
  Outer insulation: 260 °C
  Inner insulation: 320 °C
  Reduction in heat loss: 0.38 MMkcal/h
- *Saving of fuel* About 288 TPA naphtha is being saved after implementing the insulation system.
- *Economics* Total saving achieved per annum is Rs 17.6 million.

CASE STUDY AT A GLANCE

- The 3 MTPA (million tonnes per annum) NRL (Numaligarh Refinery Ltd) produces LPG (liquefied petroleum gas), naphtha, motor spirit, aviation turbine fuel, superior kerosene oil, high speed diesel, raw petroleum coke, calcined petroleum coke, and sulphur.
- Earlier, the Refinery effluent was discharged into the River Dhansiri after making sure that it met MINAS (Minimum National Standards).
- Before disposal, it used to get mixed with the effluent from the STP (sewage treatment plant) at township.
- As a water conservation measure, treated effluent was proposed to be reused instead of disposing it of to the river.
- Some modifications have been done inside the Refinery premises to reuse the effluent as fire water make up, filter blowdown in the cooling tower, and for watering saplings in the green belt zones.
- Zero discharge, or 100% recycling of treated effluent from the Refinery, has been achieved.
- Significant heat loss was taking place from the bare portion of the reformer tubes in the hydrogen plant of the Refinery.
- It was not possible to provide conventional insulation over the exposed length of the tube, as thermal expansion due to high temperature conditions causes variation in the exposed length.
- Telescopic seals with high insulation properties were provided.
- This telescopic insulation resulted in the saving of considerable amount of heat (0.38 MMkcal) and fuel (naphtha – 288 TPA).
- Total saving achieved per annum is Rs 17.6 million.

CASE STUDY 3

Larsen & Toubro Ltd

About the company

L&T (Larsen & Toubro Ltd) is India's largest engineering and construction conglomerate, with additional interests in electricals, electronics, and information technology. A strong customer-focused approach and constant quest for top-class quality have enabled L&T to attain and sustain leadership position for over six decades. L&T enjoys a premier brand image in India and its international presence is on the rise, with a global spread of over 30 offices and joint ventures with world leaders.

L&T, Powai Campus, is situated at village Tungarli, Powai, Mumbai, on approximately 95 acres of sprawling campus. The campus is divided into Powai West and Powai East. Tractors Engineers Ltd, Ewac Alloys Ltd, and L&T Infotech are also part of the campus.

The manufacturing activities at Powai Campus include mechanical fabrication for heat exchangers and other equipment required for captive projects like petrochemicals, petrol dispensing units, low-voltage switchgear products, undercarriage parts, and welding rods. The company also provides IT-related services.

Environmental profile

The company has a corporate Environment, Health, and Safety Policy. The policy stresses on the need to conserve resources like water, fuel, and electricity, minimize the use of hazardous substances, and reduce air and water pollution. The EMS (Environmental Management System) of the operating divisions of the company is ISO 14001: 2004 certified. The EMS ensures that all the processes and operations of the company meet the highest environmental standards. The core element of the EMS is environment protection and conservation.

Air pollution and control

The facility emits SPM (suspended particulate matter), SO_x (oxides of sulphur), and NO_x (oxides of nitrogen) into the atmosphere. SPM and SO_x are emitted from two furnaces, which are provided with 9.5 m tall stacks to ensure effective dispersion of the pollutants. Earlier, the fuel used in these furnaces was diesel, which was responsible for the high emissions of SPM and SO_x. Because of its polluting nature, diesel has been replaced with PNG (piped natural gas), which is relatively cleaner. Also, to ensure pollution-free cutting operation, Kjellberg under water plasma cutting system has been installed.

Water pollution and control

The effluent released from the facility is characterized by acidic pH, TSS (total suspended solids), TDS (total dissolved solids), and heavy metals. The effluent is treated in the ETP (effluent treatment plant), which is operated batchwise. The treated effluent is reused for horticultural purpose. The dried sludge is disposed of in the common hazardous waste disposal facility. On average, 4 m^3 of effluent is treated per day. The treated and untreated effluent is analysed regularly to ensure that it meets the environmental standards. Domestic effluent is treated in the STP (sewage treatment plant).

Hazardous waste

Hazardous waste generated from the facility includes coolants, used oil, waste paint residue, off-specification and discarded products, ETP sludge, and laboratory waste. Coolants and used oil are sold to authorized reprocessors and recyclers. Off-specification and discarded products and paint residue are incinerated, while ETP sludge is landfilled at Common Hazardous Waste Storage and Disposal Facility. The laboratory waste is sent to the Orian Laboratory for treatment and disposal.

Other waste

The facility also generates ferrous, non-ferrous, maintenance, and electrical scrap, welding waste, along with canteen and garden waste. The scrap is sold to approved scrap recyclers/ traders. Garden waste is composted within the premises, while food waste is collected and disposed of by a BMC (Brihanmumbai Municipal Corporation)-approved contractor.

Case study

Waste water management in Powai (West) Campus through SAFF technology based sewage treatment and reuse plant

Introduction

The Powai (West) Campus gets water from the BMC and Powai Lake. The lake water is unfit for human consumption. It is used for gardening, flushing toilets, cooling water application, in fire hydrant tanks, and other such secondary purposes. The ever increasing demand of water in the area was exerting pressure on the lake, with the BMC also expressing its inability to provide additional water.

To ease the pressure off the lake and become self-sufficient, the company decided to treat and reuse the sewage generated in the campus by setting up an STP in the campus (Figure 1). The treated sewage is required to meet the following two standards, as per the Maharashtra Pollution Control Board.

- BOD (biochemical oxygen demand) should be less than or equal to 100 mg/l.
- TSS should be less than or equal to 200 mg/l.

The system operates on the 'zero discharge' concept. It comprises primary, secondary, and tertiary units, as described in the following sections (Figure 2).

Primary treatment

The primary stage of the STP consists of screens and equalization tank fitted with diffused aeration system to equalize the organic and hydraulic load, especially from the canteen and offices. The equalization tank absorbs the wide fluctuation of the sewerage generation during peak hours and ensures uniform and continuous sewage flow to downstream treatment units. Diffusers fitted with air blowers prevent the suspended solids from settling in the tank.

Figure 1
The Larsen & Toubro sewage treatment and reuse plant

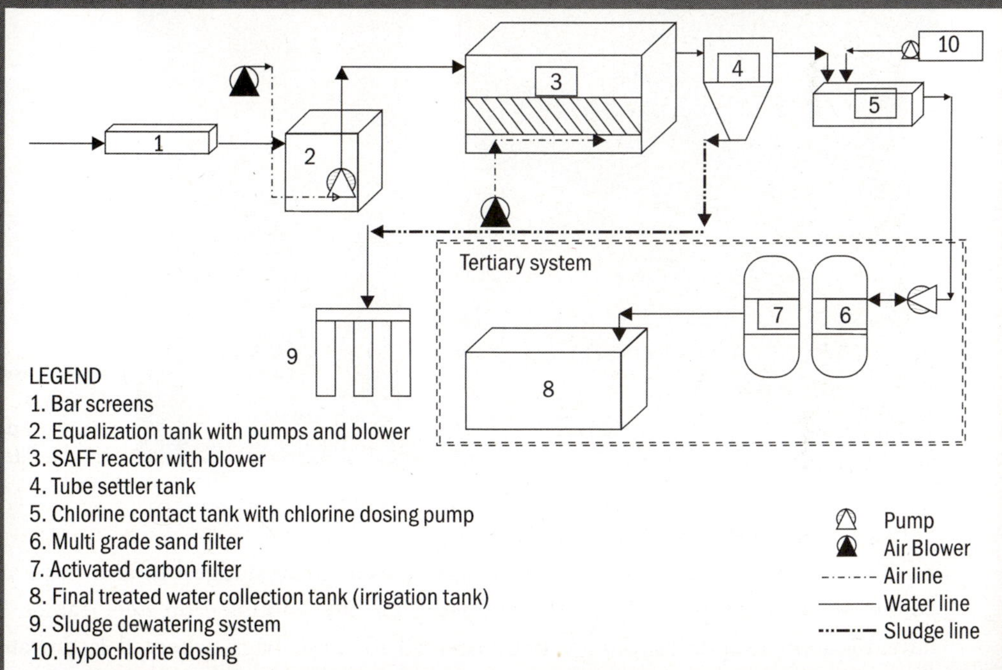

Figure 2 A diagram showing the sewage treatment and reuse plant

Secondary biological treatment

The sewage from the equalization tank is pumped by submersible feed pumps to the SAFF (submerged attached fixed film) tank (attached growth bioreactor), where organic pollutants are biodegraded. These reactors work on the attached growth principle, wherein biomass is cultivated on the synthetic media. Aerobic conditions are maintained in the bioreactors.

Gravity overflow from the attached growth bioreactors is collected in the tube settler tank (lamella clarifier). In this settling tank, sludge generated in the SAFF settles under gravity. Clear suspension from the settling tank flows under gravity to a chlorine contact tank. The settled sludge is then passed through mechanical sludge dewatering system. The sludge so generated is utilized in the existing composting plant.

In the chlorine contact tank, the treated waste water from the lamella clarifier is dosed with sodium hypochlorite for disinfection.

Tertiary treatment

The treated sewage from the chlorine contact tank is passed through multi-grade sand media filters, followed by activated carbon filter, where it is further polished (Figure 3). Filtered water is collected in the irrigation water tank, from where it is pumped to the existing hill top overhead tanks (mushroom tanks) of 250 000 litre capacity. Backwashed water from sand filters and sludge dewatering unit goes back to the equalization tank.

Figure 3
Multi-grade sand media filters

Sludge dewatering system—KOBAG

KOBAG sludge dewatering system not only dewaters the sludge but also bags and packs it continuously. It comprises specially designed filter bags mounted on steel frames. These bags have filtration properties. Sludge is transferred to these bags, where it is dewatered. Once the bags are full of solid, dry sludge, these are sealed and removed. These bags are stored in the open. The sludge volume and weight reduce during this period and dry solid content of 15%–20% is achieved.

Treated sewage

The STP treats 500 m^3 of sewage per day. The treated sewage is used for gardening, toilet flushing, and in fire hydrant tank. There is a proposal to use this water for cooling tower application as well.

Salient features of the STP

- Eco-friendly
- No bad odour
- High treatment efficiency
- Simple operation
- No heavy conventional equipment
- Needs less area
- Needs minimum maintenance
- Capacity to handle shock loads
- Reusable treated water
- Minimum sludge generation
- Dried sludge can be used as fertilizer

Cost–benefit analysis

Cost of the STP: Rs 8 745 000

Savings due to the reuse of treated sewage: Rs 2 100 000/year

Savings due to doing away with the annual cleaning of septic tanks (20 nos@Rs 15 000 per tank): Rs 300 000

Payback period: 3.6 years

Benefits

- *Water conservation* Reusing treated sewage for secondary purposes, such as gardening, toilet flushing, and cooling tower application, leads to the conservation of approximately 350 m^3/day water.
- *Savings* Annual savings to the tune of Rs 2 400 000 are achieved after the installation of the STP.

CASE STUDY AT A GLANCE

- L&T (Larsen & Toubro Ltd) is India's largest engineering and construction conglomerate, with additional interests in electricals, electronics, and IT.
- L&T, Powai Campus, is situated at village Tungarli, Powai, Mumbai.
- The manufacturing activities at Powai Campus include mechanical fabrication for heat exchangers and other equipment required for captive projects like petrochemicals, petrol dispensing units, low-voltage switchgear products, undercarriage parts, and welding rods.
- To ease the pressure off the natural water resource and meet the water demand for secondary purposes, the company decided to treat and reuse the sewage generated in the campus by setting up an STP (sewage treatment plant) in the campus.
- The system operates on the 'zero discharge' concept and comprises primary, secondary, and tertiary units, and a sludge dewatering system.
- The STP treats 500 m^3 of sewage per day. The treated sewage is used for gardening, toilet flushing, and in fire hydrant tank. There is a proposal to use this water for cooling tower application as well.
- Reusing treated sewage for secondary purposes leads to the conservation of approximately 350 m^3/day water.
- Annual savings to the tune of Rs 2 400 000 are achieved after the installation of the STP.

CASE STUDY 4

Rashtriya Ispat Nigam Ltd

About the company

Rashtriya Ispat Nigam Ltd is the corporate entity of the Visakhapatnam Steel Plant, located in Visakhapatnam city. The company also has a blast furnace grade limestone captive mine at Jaggayapeta, a captive mine for dolomite at Madharam, and a manganese ore captive mine at Cheepurupalli. All the captive mines are located in the state of Andhra Pradesh. It also has a mining lease for river sand in the River Champavathi.

The Plant has a capacity to produce 2.656 MT (million tonnes) of saleable steel, of which 2.410 MT is finished steel (2004).

Besides procuring raw materials from the captive sources, the Steel Plant meets its iron ore requirements from Bailadilla mines of the National Mineral Development Corporation, coking coal requirements through imports and coal washeries of the Coal India Ltd, SMS grade limestone through imports, quartzite requirements through purchase, and boiler coal requirement from the coal mines of the Mahanadi Coal Ltd.

The product profile of the Plant comprises wire rods, rounds, reinforcement bars (rebars), angles, channels, beams, squares, billets, and blooms. The product profile also includes basic grade pig iron, granulated slag, coal chemicals, and other by-products. The Plant also exports power to AP Transco from its captive power plant.

The Steel Plant has many technological features, which are unique amongst the steel plants in the country. The company is a pioneer in introducing many new technologies in the country. The production of TMT rebars by tempcore process is one such example. Because of the state-of-the-art technology adopted by the Plant in all its processes, it has a very good manufacturing capability.

Environmental profile

The Environment Management Department of the Visakhapatnam Steel Plant is the nodal agency for drawing out all programmes and taking initiatives in the field of the environment. All environmental parameters related to ambient air, effluent, soil, and noise are monitored regularly by the department to ensure that these parameters are within the prescribed standards. The company's combined Environment and Occupation Health and Safety Policy clearly outlines the need for protecting the environment. It also stresses on using resources efficiently, reducing wastage, and preventing pollution while complying with all relevant legal and regulatory norms.

The Visakhapatnam Steel Plant is ISO: 9001, ISO: 14001, and ISO: 18001 certified. Its ISO: 14001 has been upgraded to ISO: 14001: 2004.

Air pollution and its control measures

The main air pollutants emitted by the facility are SPM (suspended particulate matter), RSPM (respirable suspended particulate matter), SO_2 (sulphur dioxide), NO_x (oxides of nitrogen), and fugitive dust and gases. To control these emissions, a wide array of pollution control systems comprising cyclones, scrubbers, ESPs (electrostatic precipitators), and bag filters have been installed at strategic locations. The recovered dust is recycled in the sinter plant. Additionally, six gas cleaning and recovery systems are also in place. Stacks of appropriate height have been provided to ensure effective dispersal of pollutants. Low-sulphur boiler coal (with sulphur concentration <0.5%) is used in the plant processes, which has resulted in the reduction in SO_2 emissions. Regular monitoring of the ambient air, main stacks, and work zone environment is done to ensure that the pollutants do not exceed the specified norms. Acoustic enclosures are provided for high speed compressors, turbogenerators, and turbo blowers, which generate too much of noise.

Water pollution and its control measures

The effluent released from the facility contains BOD (biochemical oxygen demand), COD (chemical oxygen demand), pH, TSS (total suspended solids), and oil and grease. Various ETPs (effluent treatment plants) and toxic spillage recycling systems are in place to treat the waste water released from the facility. Mill scales and sludges are recovered from the ETPs and recycled in the sinter plant. Toxic spillage recycling systems ensure that spillages do not enter the storm water drain. Ash slurry is sent to the ash pond (Figure 1), and the treated water is recycled.

Figure 1
Pumping of drain water to ash pond

Hazardous waste

Activated sludge, tar sludge, benzol muck, oil sludge, copper scrap, and used lead acid batteries are the major hazardous waste items generated from the facility. Activated sludge, tar sludge, benzol muck, and acid sludge are charged in coke oven batteries, where they undergo incineration and get converted to pure carbon, which is as good as the coke produced from the imported coking coal. Copper scrap is recycled or used for producing value-added steel in the steel melt shop. Waste oil is sold to the registered companies.

Other waste

Other types of industrial waste generated by the facility include blast furnace slag, LD (linz donawitz) slag, blast furnace metal dust and slag, mill scale, LD dust and sludge, waste refractories, lime fines, and other waste items like coke fractions, sludge, ore dust, and so on. Blast furnace slag and LD slag are used for various applications.

Case study

Setting a world benchmark in water consumption: the Visakhapatnam Steel Plant's success story

Introduction

Steel manufacturing is a resource-intensive activity, with every tonne of steel requiring three to four times its weight of raw material and seven to eight times its weight of water. According to the International Iron and Steel Institute, the global average water consumption by a steel manufacturing industry is 9.1 m^3/tonne of finished steel, which is equivalent to nine times its weight. On the contrary, the Visakhapatnam Steel Plant consumes only 2.29 m^3 of water for every tonne of liquid steel, which is equivalent to 2.51 m^3/tonne of the finished steel. This feat was achieved by the Plant through its sustained and continuous efforts towards identifying opportunities for improving and implementing latest technologies and water conserving measures.

Salient features of water supply and distribution system at the Visakhapatnam Steel Plant

The water system at the Plant includes the following.

- Balancing reservoir with a storage capacity of 12 million m^3 for 100 days.
- 25-km-long industrial water, fire, and drinking water pipeline network.
- Combined recirculation rate of 155 000 m^3/h, with 3600 m^3/h of make-up water addition.
- 30 pump houses, 8 treatment plants, 4 overhead tanks, 22 cooling towers, and 10 sewage pump houses.
- Sewage treatment plant and effluent treatment plant.
- Twelve recirculation systems under chemical treatment.

Conservation strategy adopted

The Plant adopted a systematic and integrated approach to reduce its water consumption. The approach involved eight phases.

1 Water audit
2 Freezing of end-user specification
3 Reuse of waste water
4 100% chemical treatment of recycled water

5 Conversion of waste water to make-up water grade
6 Enforcing technological discipline
7 Involving employees/community in water conservation
8 Rainwater harvesting

Water audit

The facility did a comprehensive water audit to assess the intake of water, evaporation and seepage losses, process consumption, and process waste and boundary discharges. This audit analysed the quantity and quality of the water lost.

Freezing of end-user specification

A detailed study of the process requirements, keeping in view the process efficiency, was carried out. This study established the bare minimum quantity and quality of the water required by the end-user for sustained production performance.

Reuse of waste water

This study involved identifying points where water discharged from one stream could be used as input water in another stream. This saved the make-up water to that stream. Table 1 gives an account of the water reused in various processes.

Table 1 Reuse of waste water

Water discharge from	As input to	Water saved (m^3/h)
Power plant CT blowdown water	Ash slurry pump house	300
Oxygen plant CT blowdown water	SMS pump house	45
Chilled water plant CT blowdown water	Rolling mills pump house	30
Filters backwash water	Clarifiers of SMS	50
Package air conditioners and sample coolers	Respective cooling towers	5
Total		430

Chemical treatment (100%)

This exercise was undertaken to improve the cycle of concentration of recycling systems from the range of 2–2.5 to 5.0–5.5. This was made possible by introducing chemical treatment in recirculation systems. Use of chemicals not only prevented scaling, corrosion, microbiological fouling, nozzle chocking, and so on, but also doubled the cycle of concentrations. This resulted in the reduction of 1500 m^3/h of blowdown water, thus saving Rs 86 million per annum, which was mainly due to the reduction in equivalent amount of process water.

Converting waste water to make-up water grade

State-of-the-art technologies have been used to conserve water. The township sewage was converted to make-up water grade by using ultrafiltration (Figure 2). Another initiative involved recirculating secondary spray water in the steel melt shop by using high-rate deep bed filtration. Together, these initiatives helped in the reduction of 500 m^3/h of water.

Figure 2
Ultrafiltration plant for treating sewage

Enforcing technological discipline

The various measures taken to enforce technological measures are as follows.

- Repairing cooling tower drift eliminators.
- Strict monitoring of chemical treatment regime and working of heat exchangers.
- Repairing of U/G pipes, fixing of float valves, installing press-type valves instead of caps in canteens, placing orifice plates in pipelines to regulate the flow of water to canteen, filter, and so on.
- Strengthening of annunciation systems to provide instantaneous information in PH water tank system.
- Introducing automation in remote areas.
- Training employees on water conservation and waste management issues.

Involving employees/community in water conservation

Following measures were taken to involve employees/community in water conservation.

- Appealing through recorded voice in telephones.
- Using ticker in Cable TV.
- Distributing pamphlets containing conservation-related matter.
- Organizing public/community meetings.
- Delivering special lectures on water conservation in schools.
- Forming special inspection groups for various departments.
- Identifying the water wastage points and presenting them in public meetings.

Rainwater harvesting

The Steel Plant has installed 122 water harvesting systems (from 2002–06), which include 94 percolation pits, 1 earthen check dam (Figure 3), 9 masonry check dams, and 18 recharge wells. The indicator of the success of these water conservation measures is the industrial water consumption per year, as reported in Table 2.

Figure 3
Check dam for rainwater harvesting

Table 2 Total industrial water consumption

Year	Liquid steel production ('000 tonnes)	Total industrial water consumption ('000 m^3)	Specific water consumption (m^3/tls)	Percentage reduction over previous year	Percentage reduction over 2002/03*
2001/02	3083	15 045	4.88	–	–
2002/03	3357	13 294	3.96	18.85	18.85
2003/04	3508	11 612	3.31	16.41	32.17
2004/05	3560	9824	2.76	16.63	43.45
2005/06	3603	9224	2.56	7.23	47.54
2006/07	3606	8258	2.29	10.55	53.07

* Beginning of water conservation efforts

Cost–benefit analysis

Reduction in make-up water consumption: 1190 m^3/h

Annual savings: Rs 95.3 million

Saving in energy by way of reduced pumping hours: Rs 29.8 million

Investment made in implementing water conservation measures: Rs 134.9 million

Benefits

- *Conservation of water* There has been a 53.07% reduction in the industrial water consumption from 2002/03, the year when the water conservation measures were implemented.
- *Savings* Savings envisaged per year due to the implementation of water conservation measures are Rs 125.1 million.

CASE STUDY AT A GLANCE

- Rashtriya Ispat Nigam Ltd is the corporate entity of the Visakhapatnam Steel Plant, located in Visakhapatnam city.
- The Plant has a capacity to produce 2.656 MT of saleable steel, of which 2.410 MT is finished steel (2004).
- According to the International Iron and Steel Institute, the global average water consumption by a steel manufacturing industry is 9.1 m^3/tonne of finished steel, which is equivalent to nine times its weight.
- On the contrary, the Visakhapatnam Steel Plant consumes only 2.29 m^3 of water for every tonne of liquid steel, which is equivalent to 2.51 m^3/tonne of the finished steel.
- This feat was achieved by the Plant through its sustained and continuous efforts towards identifying opportunities for improving and implementing latest technologies and water conserving measures.
- There has been a 53.07% reduction in the industrial water consumption from 2002/03, the year when the water conservation measures were implemented.
- Savings envisaged per year due to the implementation of water conservation measures are Rs 125.1 million.

SECTION

II

Waste management and resource conservation

The corporate sector representing the manufacturing and service industries is the engine of economic growth for the country. In its pursuit of achieving a developed country status by 2020, India needs to maintain its present high GDP (gross domestic product) growth rate of 7%–8%. The Indian industry sector has a key role to play in realizing this dream. Maintenance of a high GDP necessitates increased usage of natural resources such as energy and water. The increased resource usage compounds the problem of emissions and discharge of pollutants to air, water, and land.

The 2005 MA (Millennium Ecosystem Assessment), the largest and most comprehensive multi-stakeholder review of ecosystems and their services, concluded that some two-thirds of the assessed ecosystems and their services were being degraded or used unsustainably. The rate of ecosystem degradation has been increasing rapidly for the last 50 years, compared to the 150 years prior to this.

The country, as per the estimates of the Ministry of Environment and Forests, generates around 8.4 MT (million tonnes) of hazardous waste and around 150 MT of non-hazardous waste from industrial activities. Bulk of the non-hazardous waste comprises fly ash generated from thermal power plants across the country. Only 30%–35% of the non-hazardous waste streams get recycled or reused, the balance of it requires huge land area for disposal. Therefore, there is a strong need for the country to achieve higher growth rate with optimal use of resources and adopt better management practices to reduce the generation of waste.

Despite the current economic recession, uncertain economic growth, natural disasters, rising food prices, water shortages, ongoing deforestation and biodiversity loss, local pollution concerns, and global climate change concerns, the Indian corporate sector has shown remarkable resilience and ability to rise above adversities.

Hazardous wastes from industrial sectors contain heavy metals, cyanides, pesticides, complex aromatic compounds, and other chemicals that are toxic to humans, plants or animals; are flammable, corrosive or explosive; or have high chemical reactivity. The chemical constituents of the waste may be released as air pollutants, if they are volatile, and/or may have the potential to contaminate soil and water resources (surface and groundwater), if they are water soluble. These chemicals are toxic to flora and fauna, including humans, when inhaled or ingested.

The most effective way to deal with the adverse environmental impacts of hazardous constituents of waste is to form strategies and management plans directed towards their reduction in waste streams let out from industries. Effective ways to manage the hazardous constituents of the waste and to reduce their impact on the environment are (1) source reduction, (2) recycle/reuse, (3) thermal destruction with energy recovery, and (4) landfilling of waste, with utilization of landfill gas.

There have been numerous instances where the corporate sector has shown willingness to conserve and respect natural resources, and recycle and reuse waste, therefore reducing their overall ecological footprint. This section documents case studies from the corporate sector, highlighting the initiatives taken by the sector to achieve better waste management and natural resource conservation.

CASE STUDY 5

Nava Bharat Ventures Ltd

About the company

Founded in 1975, Nava Bharat Ventures Ltd (formerly called Nava Bharat Ferro Alloy Ltd) is a manufacturing company with diversified business interests in the production of ferro alloys, sugar, and by-products and power generation. Its annual turnover is Rs 11.28 million. Its corporate office is located in Hyderabad and manufacturing units are present at Paloncha and Samalkot in Andhra Pradesh and at Kharagprasad in Orissa.

The Paloncha facility started its first ferro alloy furnace of capacity 16.5 MVA in 1975. Later, three more furnaces were installed in 1979, 1983, and 2005, respectively. Presently, the total installed furnace capacity is 74.5 MVA, and the annual production capacity ranges from 117 000 MT to 144 000 MT, depending on the product mix.

Environmental profile

The company has weaved in environment-friendly practices in all its operations. Its vision statement reveals its concern for the environment, wherein it states, '... strive to protect the environment in all our endeavours.' The company makes continual efforts to control pollution associated with all its activities through process improvement, complying with applicable legal and other requirements. It also ensures that its processes cause minimum impact on the environment. The company has taken many initiatives to conserve resources and reduce, reuse, and recycle waste.

The Quality Management System adopted by the ferro-alloy plant was formally certified as per the International Standard ISO 9002: 1987 by the Indian Register Quality Systems, Mumbai, in 1994. The Environmental Management System adopted by the Company's Works at Paloncha was formally accredited with ISO 14001: 1996 certification in 2003 by the Indian Register Quality Systems, Mumbai.

Air pollution and its control measures

Air quality in the premises is monitored regularly, and it is ensured that all parameters are within the prescribed standards. The main air pollutants emitted from the facility are SPM (suspended particulate matter), RSPM (respirable suspended particulate matter), SO_2 (sulphur dioxide), NO_x (oxides of nitrogen), and fugitive dust. To control these emissions, the company has installed various state-of-the-art pollution control devices and switched over to environment-friendly technologies. Some of these measures are listed below.

- Pulverized fuel boilers have been replaced with fluidized/atmospheric fluidized bed boilers in the thermal power plant, which has resulted in a marked reduction in SO_2 and NO_x emissions.
- Stack emissions in the power plant are being controlled by high-efficiency ESPs (electrostatic precipitators).

- To control gas fumes and dust, a gas cleaning plant has been installed, which essentially has a bag house in which a number of filter bags are arranged vertically. The gases from the furnace are extracted and made to pass through the gas cleaning plant, where SPM is separated, and clean gases are released into the atmosphere.
- To control fugitive emissions from furnaces, the company has adopted low-hood design for one of its furnaces and has also modified the hoods of other furnaces.
- To control fugitive dust, water is sprinkled in areas that handle operations generating dust. Ventilation and dust suppression systems have been installed in the coal conveyor facility.
- A green belt has been developed in the premises, which acts as a natural sink for the pollutants generated.

Water pollution and its control measures

Effluents released from the ferro alloy plant and power plant contain pH, TDS (total dissolved solids), TSS (total suspended solids), COD (chemical oxygen demand), BOD (biochemical oxygen demand), chlorides, and sulphates.

Domestic waste water generated from the township and factory is treated in a sewage treatment plant. The sewage is collected in a collection sump and pumped into the aeration tank. Micro-organisms, in the presence of air, bring down the BOD and COD levels to the desirable limits. The suspended particulates are allowed to settle in the clarifier.

Water from these ponds is recycled/reused for various purposes like bed ash pumping in the power plant, slag granulation in the ferro alloy plant, and for maintaining the green belt. Thus, the company has managed to achieve the 'zero discharge' status.

Hazardous waste

Used/waste lubricating oil, used lead batteries, copper cables and scrap, and process dust are the different types of hazardous waste generated from the facility. All the waste items, except process dust, are sold to authorized agencies. Process dust is reused in the production of sinters, which is ultimately used in furnaces.

Other waste

Other industrial waste items generated from the facility include slag granules from the ferro alloy plant, fly ash from the power plant, bed ash from the power plant, fumes from the gas cleaning plant, and manganese ore fines from the ferro alloy plant. The slag produced in the ferro alloy furnaces is granulated and disposed of to manufacturers for making fly ash bricks. It is also used as a partial substitute for sand in concrete mixes. The fly ash is disposed of to units involved in making cement and fly ash. The bed ash is used for filling low-lying areas. The fumes collected from the gas cleaning plants connected to the ferro alloy furnaces are mixed with manganese ore fines and sintered for reusing the same in the furnaces for recovering the manganese.

Case study

Utilizing waste heat from the dust-laden flue gas of the fourth furnace for heating main condensate and feed water in the power plant

The facility in Paloncha has installed a 25 MVA furnace with low hood design to minimize fugitive emissions from the furnace floors by employing Pyromet Technology, South Africa. About 140 000 Nm^3/h of hot gases are released from the furnace at a temperature of about 380 °C. Initially, these gases were cooled to a temperature of less than 230 °C by passing them through a radiant gas cooler. Earlier, these cooled gases used to be diverted to the filter bag house (gas cleaning plant), where the fumes were collected and the dust-free gases were released into the atmosphere. Consequently, a lot of heat was going waste.

The company did a lot of R&D on tapping this heat of the dust-laden gases and using it in the 32-MW power plant for heating condensate and feed water. Towards this end, the radiant gas cooler was replaced with a cross-flow type of heat exchanger at an investment of Rs 50 million. The flue gases are passed vertically down over the bundles of horizontal water tubes. The water tube bundles are divided into HP (high pressure) and LP (low pressure) tube bundles called HPHX (high pressure heat exchangers) and LPHX (low pressure heat exchangers). Feed water and main condensate are heated in the HPHX and LPHX, respectively. The flue gases coming from the furnace pass through the HPHX first and then the LPHX, before finally entering the gas cleaning plant through an ID fan.

Cost–benefit analysis

Investment made on the waste heat recovery system: Rs 50 million
Cost of coal saved: Rs 50 000/day
Cost of power saved: Rs 3000/day
Other savings: Rs 2000/day
Total savings: Rs 55 000/day
Payback period: 909 days

Benefits

- Specific steam consumption in the power plant turbine has decreased by 7–8 TPH (tonnes per hour).
- Coal consumption has decreased by 1.5 TPH at a GCV (gross calorific value) of 3600 kcal/kg, which means that the company saves about 68 tonnes of carbon dioxide in a day, equivalent to 68 carbon credits per day.
- Total savings due to coal and power conservation are to the tune of Rs 55 000/day.
- Due to low operating temperature and differential pressure, life of the filter bags of the gas cleaning plant has increased.
- The maintenance cost of the gas cleaning plant has reduced considerably.

CASE STUDY AT A GLANCE

- Founded in 1975, Nava Bharat Ventures Ltd (formerly called Nava Bharat Ferro Alloy Ltd) is a manufacturing company with diversified business interests in the production of ferro alloys, sugar, and by-products and power generation.
- About 140 000 Nm3/h of hot gases are released from the furnace at a temperature of about 380 °C.
- Initially, these gases were cooled and routed to the gas cleaning plant, from where the dust-free gases were released into the atmosphere. Consequently, a lot of heat was going waste.
- To tap this heat for heating condensate and feed water in the 32-MW power plant, a heat exchanger was installed in place of the radiant gas cooler.
- Specific steam consumption in the power plant turbine has decreased by 7–8 TPH (tonnes per hour).
- Coal consumption has decreased by 1.5 TPH at a GCV (gross calorific value) of 3600 kcal/kg, which means that the company saves about 68 tonnes of carbon dioxide in a day, equivalent to 68 carbon credits per day.
- Total savings due to coal and power conservation are to the tune of Rs 55 000/day.

CASE STUDY 6 Sudarshan Chemical Industries Ltd

Main facility

About the company

Sudarshan Chemical Industry Ltd, established in 1968, mainly deals in pigments, pesticides, and other agro chemicals. Its facilities are located in Pune, Roha, and Mahad, all in Maharashtra. The company manufactures a full range of pigments, including conventional pigments, high performance pigments, and effect pigments. These are used in various applications such as coatings, inks, plastics, cosmetics, and so on. The total revenue for 2007/08 amounted to Rs 4023 million as against Rs 3907 million for the previous year.

The company's facility at Roha in Raigad, Maharashtra, was established in 1974 (Figure 1). Its main products include inorganic pigment, organic pigment, effect pigment, and agro chemicals.

Environmental profile

Sudarshan Chemical Industry Ltd was the first chemical company in India to be awarded the ISO 9001 Certificate, in 1991. Its Pune, Roha, and Mahad plants have also been awarded the ISO 14001 Certificate. The Roha and Mahad plants have also been certified for OHSAS 18000. The Mahad plant has been certified OHSAS 18001 by BVQI for implementing systems aimed at reducing occupational risks in work areas, besides focusing on health and safety.

The company has an environment management programme in place. This programme includes waste minimization, prevention of pollution, reuse of natural resources, greening of the earth, and so on. Canteen waste is converted into biogas, which is used as a fuel in the canteen. The company has a well-equipped environment laboratory for analysing and treating various effluents.

Air pollution and its control measures

Suspended particulate matter

To control suspended particulate matter, bag filters have been installed at appropriate locations like pulverizer, blender, and so on.

Oxides of sulphur and nitrogen

Scrubbers with caustic solution are provided to scrub oxides of sulphur and nitrogen, hydrochloric acid, chlorine gas, and volatile organic compounds released during the manufacturing process. Emissions of these pollutants are monitored regularly and reports are submitted to the MPCB (Maharashtra Pollution Control Board).

Fugitive emissions

These are controlled using process engineering design and good operating practices. Standard engineering practices like providing condensers with raw water and chilled water cooling are followed. Interlock systems are also provided for controlling fugitive emissions. Workplace monitoring is also done to keep these emissions under control.

Ammonia and chlorine

Appropriate scrubbers and absorbers are provided wherever chlorine and ammonia are generated during reactions. Leak detection and repair systems are also provided for chlorine and ammonia.

Table 1 presents the concentrations of air pollutants emitted in the premises, along with the MPCB standards.

Table 1 Air pollutants emitted from the premises

Pollutant	*Concentration of the pollutant*	*MPCB standards*
Particulate matter	12.9 mg/Nm^3	150 mg/Nm^3
Sulphur dioxide from boiler	237 kg/day	1200 kg/day
Sulphur dioxide from process	6.4 mg/Nm^3	50 mg/Nm^3
Hydrogen chloride	6.2 mg/Nm^3	35 mg/Nm^3
Oxides of nitrogen	24.5 PPM	50 PPM
Chlorine	1.1 PPM	3 PPM

MPCB – Maharashtra Pollution Control Board; PPM – parts per million

Noise pollution control measures

Acoustics have been provided for the diesel generation sets and high noise making compressors. Silencers are also provided for compressors. Protective gears like ear muffs are provided to workers working in noisy areas.

Water pollution and its control measures

BOD (biochemical oxygen demand), COD (chemical oxygen demand), oil and grease, total dissolved solids, total suspended solids, and heavy metals present in the effluent are controlled by treating the effluent in the ETP (effluent treatment plant) using neutralization and flocculation processes, which was installed with an initial cost of Rs 35 million.

The ETP laboratory is equipped with a pH meter, DO (dissolved oxygen) meter, COD digester, BOD incubator, oven, and furnace. Analysis of parameters like pH, total suspended solids, total dissolved solids, BOD, COD, mixed liquor suspended solids, mixed liquor volatile suspended solids, and so on is done daily. Hourly sampling is done to analyse parameters. Apart from this, recycling of water in the dye intermediate and pigment plants saves about 500 litres of water daily, resulting in a saving of Rs 900 000 per annum. Table 2 gives an account of the water pollutants generated, along with their concentrations and MPCB standards.

Table 2 Concentration of select water pollutants in the effluent

Pollutant	*Concentration of the pollutant in the discharge*	*MPCB standard*
pH	7.1	5.5–9
Suspended solids	40.8	100
BOD (3 days)	28.8	100
COD	145.4	250
Oil and grease	1.58	10
Total ammonical nitrogen	15.7	50
Lead	Below detection limit	0.1
Zinc	0.19	5
Cadmium	0	2
Phosphates	0.17	0.5
Pesticides	Nil	0.1

MPCB – Maharashtra Pollution Control Board; BOD – biochemical oxygen demand; COD – chemical oxygen demand

Note All parameters, except pH, in mg/L.

Case study

To recover heavy metal cadmium from the effluent generated at the cadmium production unit

Cadmium: the hazardous waste

One of the pigment products is made from cadmium. During the manufacturing process, the effluent released contains cadmium in both soluble and insoluble forms. From 2004, a process was put in place to reduce the concentration of this heavy metal. Gypsum is used for flocculating and precipitating cadmium from the effluent, and the gypsum sludge is sent to secured landfills.

Method used for treating effluent

For treating the effluent, filter presses, tanks, and pumps are employed. Trained manpower is used, but no extra manpower was deployed for this purpose. A pre-designed process for the recovery of cadmium is used. The reduction in cadmium concentration is achieved by waste recycling method (Environment Management – ISO 14001).

Process for effluent treatment

Effluent containing cadmium is treated, filtered, and reprocessed before it is put back in the process. The effluent is collected in an effluent collection tank by gravity. The effluent thus collected is then neutralized with lime. It is then subjected to flocculation. Cadmium is made to precipitate out along with gypsum. Neutralized and flocculated effluent is then taken to the primary clarifier, from where underflow is taken to the filter press to separate out the gypsum sludge. Overflow from the clarifier is taken to the secondary treatment plant via equalization tank. The treated effluent is then taken to the secondary clarifier, from where the settled biomass is recycled back to the aeration tank. The treated effluent thus obtained is relatively free of cadmium. The gypsum sludge is sent for disposal to secured landfills. Figure 2 shows the process of treating effluent containing cadmium, and figure 3 shows the treatment plant for treating effluents containing cadmium.

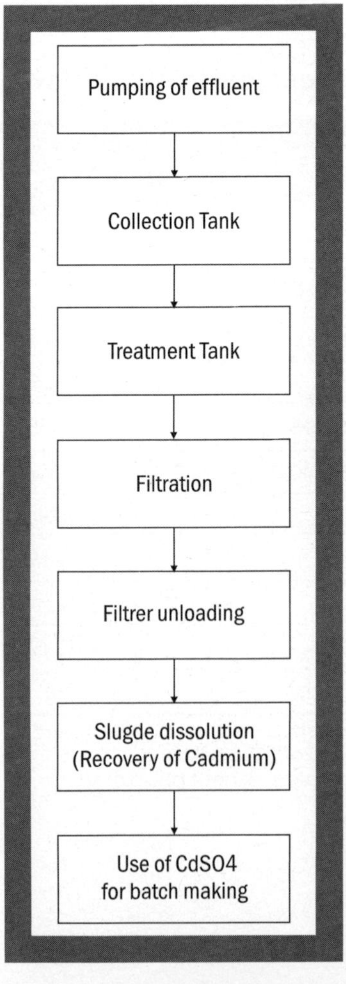

Figure 2
Flow chart showing the process of treating effluent containing cadmium

Figure 3
Treatment plant for treating effluents containing cadmium

Benefits

- Cadmium content in gypsum sludge is reduced from 800 PPM (parts per million) to <50 PPM.
- Cadmium process sludge is reduced to a great extent.
- Cadmium is an expensive metal, hence, its recovery results in cost benefit. Recovery of cadmium
- from the waste results in a net saving of Rs 600 000 per annum.
- The hazardous waste is disposed of in a scientific way.
- There is an increased awareness in employees towards the environment.

CASE STUDY AT A GLANCE

- Sudarshan Chemical Industries mainly deals in pigments and agro chemicals.
- The annual turnover for 2007/08 was Rs 4023 million.
- From 2004, a process was put in place to reduce the concentration of cadmium in the effluent by recovering the same.
- Cadmium content in gypsum sludge was reduced from 800 PPM (parts per million) to <50 PPM.
- Recovery of cadmium from the waste results in a net saving of Rs 600 000 per annum.

CASE STUDY 7 ACC Ltd

About the company

ACC Ltd is India's foremost manufacturer of cement and concrete. Its operations are spread throughout the country, which is dotted with 14 modern cement factories, 19 sales offices, and several zonal offices. In 70 years of its existence, ACC has been a pioneer in the manufacture of cement and concrete, and a trendsetter in many areas of cement and concrete technology, including improvements in raw material utilization, process improvement, energy conservation, and development of high concrete.

ACC Kymore Works is located at the foot of Kymore hills in Katni district in Central Madhya Pradesh. The facility has an installed capacity of 90 000 tonnes/annum. The plant has 3.5 MTPA (million tonnes per annum) clinkering capacity and 2.2 MTPA grinding/packing capacity.

The major products manufactured are ordinary Portland cement and Pozzolona Portland cement. Annual production of these products is 0.2 MT (million tonnes) and 2 MT, respectively. The annual turnover of the company in 2007/08 was Rs 71 359.7 million.

Environmental profile

ACC's vision highlights the company's commitment to sustainable development. The vision statement underlines the company's commitment to environment protection. The company has adopted latest technologies to protect and upgrade the environment around the units and conserve and protect all natural resources.

The Environment Management Organization has been strengthened at the corporate as well as plant level. The Environment and Energy Conservation Cell was formed in September 2007 with a clear road map to pursue emission and energy reduction. The ACC plant is ISO 14001 certified.

Awards/accolades

- Golden Peacock Eco-innovation Award 2008 by the World Environment Foundation.
- Greentech Environment Excellence Silver Award in 2008 by the Greentech Foundation.

Air pollution

The main air pollutants released from the facility are SPM (suspended particulate matter), RSPM (respirable suspended particulate matter), SO_2 (sulphur dioxide), NO_x (oxides of nitrogen), and fugitive dust and gases.

Air pollution control measures

SPM and RSPM

The SPM and RSPM comprising clinker fines, cement dust, raw material fines, and limestone fines are controlled by ESPs (electrostatic precipitators) and bag houses that have been installed at strategic locations. Transfer point dust collectors are also installed at all the material transfer points. Pollution control equipment are interlocked with main pollution control devices. Raw material fines, clinker fines, and cement fines are reused. Stacks of adequate height are provided. Mobile road sweepers regularly run on plant roads. Water tankers spray water in plants and mines. A green belt has been developed in and around the factory premises.

SO_2

Limestone is used as a raw material for manufacturing cement. It is the best scrubber for SO_2. Hence, the quantity of SO_2 vented through the kiln stack is very negligible, sometime nil.

NO_x

Multichannel, low-NO_x burner has been installed in the kiln to control NO_x emissions.

Fugitive emissions

Insertable bag filters have been installed at each transfer point of the material handling system. The discharge points of the chutes of the conveying equipment are designed for better control of fugitive emissions. Nine dust extraction systems and five dust suppression systems have been put in place in captive power plants.

Noise

The machines are well enclosed to reduce noise. A greenbelt has also been developed to attenuate noise levels.

Water pollution

The water pollutants present in the waste water generated from the facility comprise BOD (biochemical oxygen demand), COD (chemical oxygen demand), pH, oil and grease, thermal discharge, TSS (total suspended solids), TDS (total dissolved solids), and heavy metals.

Water pollution control measures

The Kymore Cement Works has installed an ETP (effluent treatment plant) for the captive power plant for effluents generated from various sources like cooling tower blowdown and for compressor cooling water and treated DM (demineralized) plant effluent. The water after minimum treatment is reutilized in clinker-making units, in the cement plant in gas cooling tower, and in cooling and dust suppression system, resulting in the conservation of natural resources and lowering of the production cost. The ETP was installed with a total cost of Rs 20 million.

Hazardous waste

Used oil and oil-soaked cotton waste are the different types of hazardous waste generated. The used oil is sold to the CPCB (Central Pollution Control Board)/MoEF (Ministry of Environment and Forests)-authorized recyclers, and oil-soaked cotton waste is burnt in the kiln furnace.

Case study

Co-processing of expired finished products from Hindustan Unilever Ltd at ACC Kymore Works

Introduction of the technology

The concept of co-processing technology was introduced in ACC in 2005 by its promoter Holcim (Switzerland). HUL (Hindustan Unilever Ltd) was ACC's first customer in terms of implementing this technology. HUL is in the business of manufacturing fast moving consumer goods such as shampoos, lotions, creams, talcum powder, and so on, which have definite shelf-life. Once their shelf-life gets over, these are not fit for use. In addition, sometimes, there are damaged lots that cannot be put back into the market. These products need proper management. The facility at Kymore is assisting HUL in the complete disposal of these products (Figure 1).

Figure 1
Kymore Cement Works where expired finished products from HUL are being co-processed

Conceptualization of the technology

Cement making is an energy- and material-intensive process. After limestone mining, grinding, and homogenization, the next step is calcinations of calcium carbonate, followed by sintering the resulting calcium oxide with silica, alumina, and iron oxide at a high temperature to form clinker. The clinker is then ground or milled with gypsum and other constituents to produce cement. Producing 1 tonne of clinker requires, on an average, 1.5–1.6 tonnes of raw materials. Traditionally, the primary fuel used has been coal. But these days, various types of waste materials are used as fuel. These waste materials are co-processed in a rotary kiln with multi-stage suspension pre-heater and pre-calciner.

Co-processing refers to the use of waste materials in industrial processes such as cement, lime or steel production or any other large combustion plants. It means the substitution of primary fuel and raw material with waste, leading to the recovery of energy and material from waste. Waste materials used for co-processing are called AFR (alternative fuels and raw materials).

The use of AFR requires the same stringent controls as those imposed on traditional fuels and raw materials. Also, proper systems, procedures, and standards, including environmental management systems, are required to be in place.

Figure 2 gives a snapshot of the co-processing process followed at the facility.

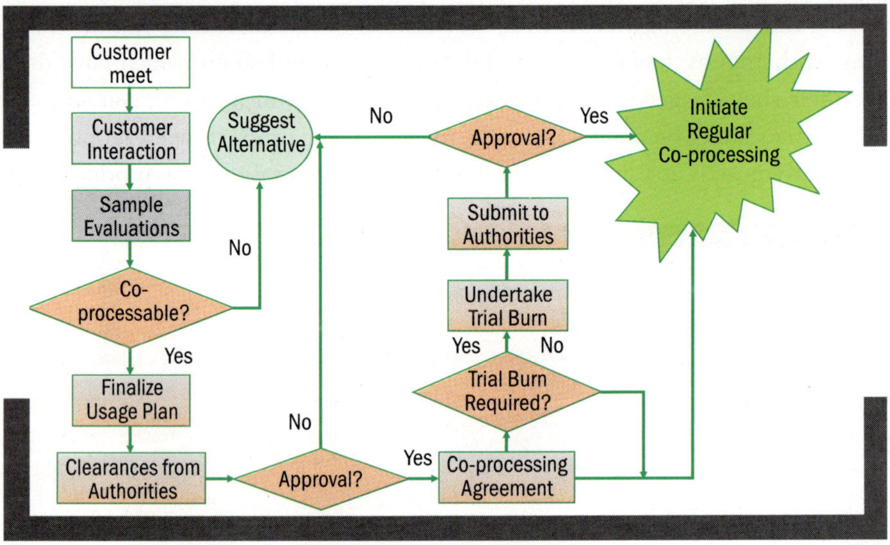

Figure 2 Procedures followed for co-processing of a waste stream at ACC Cement Works

Proactive practices and procedures followed at Kymore Works for co-processing trade rejects

The Kymore Cement Works has installed a CEMS (continuous emissions monitoring system), which monitors (online) not just dust, but also CO_2 (carbon dioxide), CO (carbon monoxide), NH_3 (ammonia), SO_2, NO (nitric oxide), VOC (volatile organic compounds), and so on. Minimal emissions from AFR are ensured.

Volatile AFR is fed via the cold end of the kiln system, where temperatures are not adequate for their destruction. The concentration of volatile heavy metals and input of other heavy metals were limited in the AFR by determining the adequate feed rates by fuel mix optimizer model. The feed rate of the trade rejects in the calciner feeding system, mostly in the form of foil, was much below the potential TSR (thermal substitution rate).

Checking negative impacts

The negative impacts of co-processing the trade rejects in the kiln on clinker, cement, and concrete quality are checked using following ways.

- Carefully monitoring if the sulphurization degree in the clinker becomes too low by increased sulphur volatilization.
- Strictly limiting the input of water of each AFR to avoid production loss.
- Strictly observing the limits for kiln feed fineness to avoid degradation of clinker burnability and clinker variability.
- Strictly limiting the input of some critical minor elements such as phosphorus pentaoxide, which affects the setting time of cement.

Procedures followed prior to accepting trade rejects from HUL

Preliminary screening of the waste

The waste material is screened for any hazardous material like anatomical hospital waste, asbestos-containing waste, bio-hazardous waste, explosives, radioactive waste, and so on.

Industry survey

Details like the rate of generation of trade rejects and the process of waste generation are obtained from the industry, which is a part of the risk assessment procedure.

Waste profile

Details of the waste profile are obtained to quantify risk from the rejects and to take precautions for handling and storing these rejects. Representative samples collected from the HUL warehouse are sent to ACC R&D (research and development) Centre.

AFR technical assessment

Thermal characteristics of the kiln system are assessed, and the point of feeding of the trade rejects in the kiln system is established. It is ensured that process sulphur/alkali/chloride balance is within the permitted range for stable kiln operation. This assessment helps in addressing the problems of production loss and quality, and additional problems from the kiln stack. The assessment has also shown that a maximum of 23% of the total thermal requirement in the plant may be met by feeding trade rejects and other wastes.

Baseline emission monitoring

The baseline monitoring of the stack at kiln no. 1, when no waste or AFR materials are used, formed the basis of incremental emissions during co-processing of waste in the subsequent years.

Commercial proposal, negotiation, and legal agreement

The commercial terms and conditions are finalized on the basis of a nominal value of the services being offered by ACC to HUL. HUL is responsible for transporting trade rejects from the HUL warehouses to ACC Kymore Cement Works.

Hazard identification, risk assessment, and control

ACC and HUL jointly developed a risk assessment procedure and formulated action points for implementing necessary controls.

Transportation and receipt of trade rejects

A pre-delivery inspection is done to ensure that only the agreed material is delivered. A delivery schedule is worked out. The AFR coordinator in the cement plant conducts an inspection of the delivery, and the truck is weighed.

Protocols for sampling and testing of the delivered AFR

The parameters for delivery acceptance criteria are analysed at the AFR laboratory. The cement company accepts the consignment only if the variation in trade rejects is within the range as agreed between the parties.

Protocols prior to feeding of AFR

The feed rate of the AFR is established during the trial run, wherein the material fed to the kiln is optimized, and emissions and other operational parameters are closely monitored. The feeding of the AFR may start only after all safety precautions are in place and workers handling the materials are well trained.

Protocols for feeding of AFR

Before feeding trade rejects into the kiln, kiln-stable conditions, in terms of operation, are ensured. Emission records are regularly collated. Clinker samples are drawn and analysed frequently.

Protocols for verification and monitoring

Emissions data is monitored and recorded regularly. Elaborate baseline monitoring is done as per Holcim EMR (Environmental Monitoring and Reporting) guidelines by a CPCB-approved, ISO 17025 certified third party monitoring agency.

Feeding system modification for process improvement

For co-processing waste materials in cement kilns, a starter kit feeding system was installed at the calciner, which consists of a hoist machine, hopper, belt feeder, double flap damper, and slide gate. The hoist machine carries the material from the ground level to the feeding platform at pre-heater tower, where belt feeder has been installed. The hoist carries 3–4 tonnes of materials in an hour. The material is unloaded from the bucket on the feeding platform. The waste materials are introduced into the initial opening of chute through a belt conveyor. A double flap damper is used in order to reduce the false air entry, and a slide gate has been incorporated in the system, keeping safety considerations in view. Both are interlocked with kiln operation, ensuring safety in case of backfiring.

Results

The project is carried out in three phases: pre-co-processing, co-processing, and post-co-processing. Stacks are generally monitored, and monitoring results of one such trial are given in Table 1. In the table, the incremental variations in emissions during co-processing and post-co-processing vis-à-vis baseline measurements are compared with the CPCB norms for common hazardous waste incinerators. The CPCB guideline provides emission standards for

Table 1 Stack monitoring results

Parameter	Unit	Norm	Measured stack emission during the trial Pre-co-processing (A)	Co-processing (B)	Post-co-processing (C)	Change in emissions during co-processing of the waste (B–A)	Change in emissions after co-processing of the waste (C–A)
Dioxin and furan	Ng TEQ/ Nm^3	0.1	0.016	0.006	0.005	–0.01	–0.01
Total organic carbon	mg/Nm^3	20	5.92	5.29	6.34	–0.63	0.42
Benzene	mg/Nm^3	–	0	0	0	0	0
Hydrochloric acid	mg/Nm^3	50	4.62	6.07	5.12	1.45	0.5
Hydrofluoric acid	mg/Nm^3	4	0.047	0.053	0	0.0057	–0.047
Sulphur dioxide	mg/Nm^3	200	64	70.66	35	6.66	–29
Suspended particulate matter	mg/Nm^3		90.086	66.65	71.66	–23.43	–18.42
Carbon monoxide	mg/Nm^3	100	305.33	452	406.66	146.66	101.33
Oxides of nitrogen	mg/Nm^3	400	665	695.6	701.66	30.66	36.667
Ammonia	mg/Nm^3	–	0	0	0	0	0
Mercury	mg/Nm^3	0.05	0.02	0.03	0.05	0.01	0.025

each parameter during the incineration of waste. The analysis results of the stack parameters revealed that the emission values are well below the standards set by the CPCB.

Conclusion

The emissions are generally independent of the usage of rejects in the cement plant. The cement making process requires high temperature in the kiln (1400–1450 °C) and a high residence time of 4–5 seconds. Such high temperature conditions ensure that no noxious emissions take place during the co-processing of the waste (Figures 3–5).

Benefits

The environmental benefits are as follows.

- Using waste as fuel and raw material in cement kilns through co-processing helps in its management in an environmentally sustainable manner while conserving the traditional raw materials and fuels for the future generations.

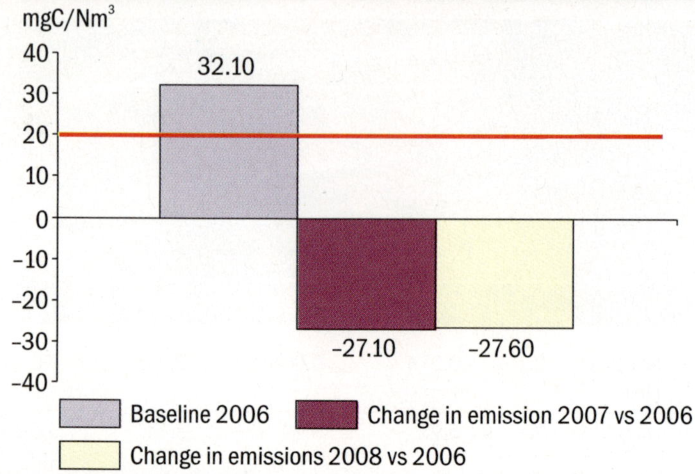

Figure 3 Total organic carbon emissions in 2007/08 compared with baseline emissions (2006)

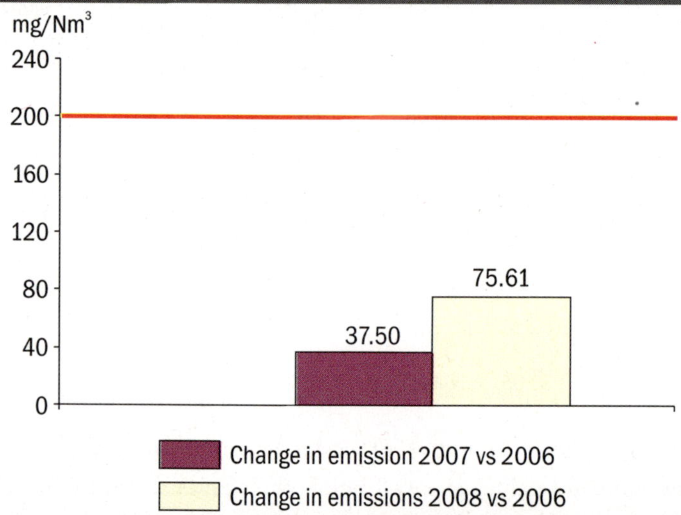

Figure 4 Total sulphur dioxide emissions in 2007/08 compared with baseline emissions (2006)

- Substituting fossil fuel and virgin raw material with waste reduces extraction from quarries and mines, and reduces CO_2 emissions.
- Co-processing leaves no residue, as the incombustible content of the waste materials is incorporated in the clinker matrix.
- There is reduction in the overall emissions. The emissions are found to be independent of the usage of materials in the cement process.

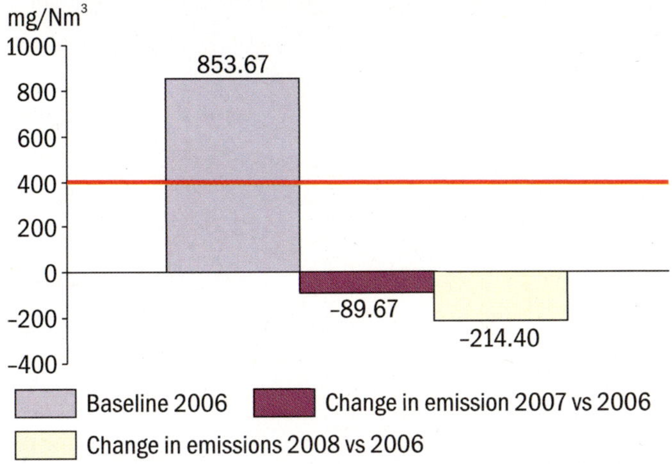

Figure 5 Total emissions of oxides of nitrogen in 2007/08 compared with baseline emissions (2006)

CASE STUDY AT A GLANCE

- ACC Ltd is India's foremost manufacturer of cement and concrete.
- The annual turnover of the company in 2007/08 was Rs 71 359.7 million.
- The concept of co-processing technology was introduced in ACC in 2005 by its promoter Holcim (Switzerland).
- The facility at Kymore is facilitating complete disposal of the waste products generated by Hindustan Unilever Ltd, through co-processing.
- Use of waste as fuel and raw material in cement kilns through co-processing helps in the management of waste in an environmentally sustainable manner while conserving the traditional raw materials and fuels for the future generations.
- There is a reduction in the overall emissions. The emissions are found to be independent of the usage of materials in the cement process.

CASE STUDY 8

Vikram Cement

Overview of Vikram Cement

About the company

Vikram Cement (a unit of Grasim Industries Ltd), a flagship company of Aditya Birla Group, is involved in the manufacture of grey cement and clinker. Aditya Birla Group is one of the largest business houses in India. The group has highly diversified business comprising more than 80 manufacturing units, financial services, and consulting and trading divisions, spread over 20 countries. It is the world's largest producer of viscose staple fibres. Its production line includes cement, caustic soda, filament yarn, flax yarn, rayon pulp, petrochemical products, copper, and aluminium.

Grasim Industries Ltd is a multi-dimensional and multi-locational industrial conglomerate producing viscose staple fibre, cement, sponge iron, rayon pulp, and so on.

Vikram Cement, the acknowledged pioneer of TPM in the Indian Subcontinent, is one of the single location highest cement manufacturing plant, with a production capacity of more than 3 MTPA (million tonnes per annum). The company, with a capacity of 0.5 MT (million tonnes), was established in 1985 at Khor, Madhya Pradesh (Figure 1). Presently, it has three production lines, that is, VC Line-1 of rated capacity 2800 TPD (tonnes per day), VC Line-2 of rated capacity 2800 TPD, and VC Line-3 of rated capacity 3800 TPD. The company's turnover in 2007/08 was 15.70 million.

Environmental profile

The company works on the concept of world class manufacturing, maintaining highest environmental standards in all its operations and processes. It is the first cement plant in India to receive ISO 14001 certification from Det Norske Veritas. Environmental impact studies are conducted periodically and based on these studies, procedures are changed or modified. The company has upgraded its Environmental Management System to ISO 14001: 2004.

Air pollution and its control measures

The major air pollutants emitted by the facility include SPM (suspended particulate matter), RSPM (respirable suspended particulate matter), SO_2 (sulphur dioxide), NO_x (oxides of nitrogen), and fugitive dust and gases. To control these emissions, the facility has installed state-of-the-art pollution control equipment like ESPs (electrostatic precipitators), bag houses, and bag filters in all its major units. Besides, a green belt has been developed in the premises. Wet drilling and water sprinkling methods are employed to control fugitive dust.

Water pollution and its control measures

The facility does not generate any waste water. Water is used only for cooling and spraying purposes, which then evaporates into the atmosphere. The domestic waste water discharged from the residential colonies is treated in a sewage treatment plant. The treated water is used in horticulture.

Hazardous waste

Furnace oil sludge and used lubrication oil are the hazardous waste items produced in the facility. These are sold to authorized recyclers and re-processors.

Case study

Initiatives to use processed municipal solid waste (refuse-derived fuel) as an alternative fuel in the cement manufacturing process

Introduction

Rapid pace of urbanization and industrialization, coupled with increasing population, has given rise to the problem of large-scale generation of MSW (municipal solid waste). Lack of infrastructure and proper planning make the task of managing this waste difficult. Statistics reveal that only about 50% of the total waste generated in the urban areas is disposed of in landfills, while the remaining is left unattended, causing environmental and health problems. Hence, there is a need to adopt new, environment-friendly technologies like producing RDF (refuse-derived fuel), to manage the MSW.

Vikram Cement's initiatives

The company, in association with the Jaipur Municipal Corporation, has set up an integrated MSW processing plant to produce RDF from MSW on 25 acres of land at village Langariwas, about 25 km from Jaipur. The plant comprises an MSW storage yard, processing machineries, fuel storage unit, and a green belt. The plant is designed to process about 500 TPD of MSW and generate about 150 TPD of RDF in the form of fluff. The technology used has been imported from Germany.

Process description

The process of conversion of MSW into RDF involves size reduction, homogenization, metal separation, segregation, further size reduction, and densification. The Jaipur Municipal Corporation supplies the MSW for the plant and ensures that the waste does not contain construction waste, hospital waste, slaughterhouse waste, and drainage silt.

In the MSW storage yard, the waste is segregated to remove large-sized particles (Figure 2). The waste is then fed into a primary shredder (Figures 3 and 4), where waste of size of up to 1000 mm is shredded to approximately 0–200 mm grain size. The shredded material is homogenized, ripped apart, and made ready for further processing. The magnetic metal parts are separated by means of a magnetic separator (Figure 5), which are collected and sold to recyclers.

Figure 2
Unloading of municipal solid waste

Figure 3
Feeding of municipal solid waste for processing

Figure 4
Feeding of municipal solid waste in the process

Figure 5
Magnetic separation

The waste stream is then separated into two fractions by passing it through a drum screen. The screened mineral and organic fractions are sent to a reject bin and the rejects are either discarded or used for making compost. The discharge product from the screen is reduced in moisture content, homogenized, and increased in calorific value before shredding to required finish product fineness.

The waste is sent to the ballistic separator, where heavier mineral fractions like stones, glass pieces, and ceramics are separated from the lighter fraction on account of difference in densities by ballistic separation. The organic light fraction is sent for further processing, and the inorganic portion of the high density material is considered unusable and is dumped into a reject bin. With the reduction in inorganic materials, calorific value of the waste increases. The light fraction is further homogenized and directed to the fine shredding machines.

In the final step of fine shredding, the high calorific value material is shredded to the final grain size of 0–50 mm (Figure 6) and the product RDF fluff is ready to be used (Figure 7). Figure 8 gives the process flow diagram of the MSW processing plant.

Figure 6
Size reduction process

Figure 7
Processed RDF on the conveyor belt

This RDF is used as a fuel in the cement manufacturing process at the Vikram Cement Line-III at the rate of 6.25 MT/h, replacing an equivalent quantity of fossil fuel.

Production and utilization of RDF from MSW is one of the best suitable options for sound MSW disposal. As it not only saves the precious fossil fuels but its co-incineration in industrial operations also reduces the landfill area and GHG (greenhouse gas) emissions resulting from

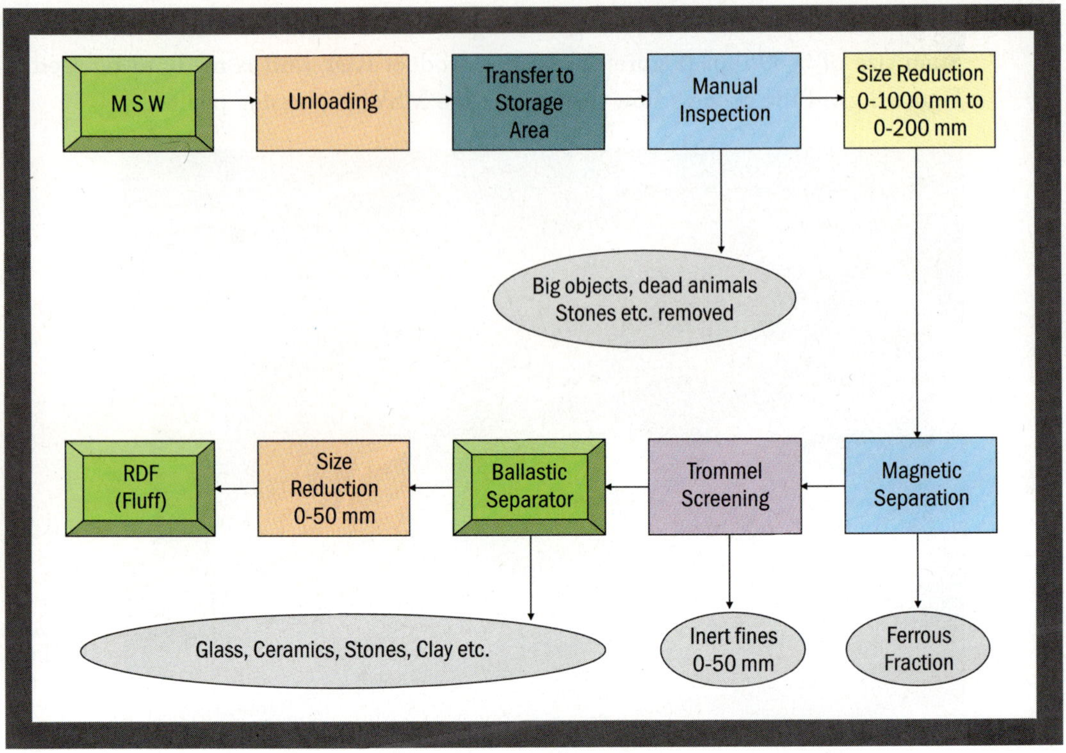

Figure 8 Flow chart describing the working of a MSW processing plant

burning of fossil fuels. During the period of July 2007–March 2008, Vikram Cement was able to reduce CO_2 emissions by 7242 MT.

Cost–benefit analysis

The company has invested approximately Rs 16 868 million to establish the MSW processing facility at Jaipur. Besides, Rs 14 099.9 million was invested to develop alternative MSW-based fuel feeding facility at the Vikram Cement plant. In 2007/08, the average cost of the alternative fuel used was Rs 1331 per GCV (gross calorific value) against Rs 978 per GCV for coal. This project has been registered as a CDM (Clean Development Mechanism) activity, which will improve its cost economics.

Benefits

- Energy potential of MSW is realized.
- Utilization of RDF has resulted in the conservation of fossil fuels. MSW is disposed of in an environment-friendly manner.
- Landfilling requirement of MSW has reduced by 60%–65%.
- There is an overall reduction in greenhouse gas emission. During the period July 2007–March 2008, the company was able to reduce CO_2 emissions by 7242 MT.
- The process of RDF production is clean, as it does not generate any residual waste that needs to be treated.

CASE STUDY AT A GLANCE

- Vikram Cement (a unit of Grasim Industries Ltd), a flagship company of Aditya Birla Group, is involved in the manufacture of grey cement and clinker.
- The company's turnover in 2007/08 was 15.70 million.
- The company, in association with the Jaipur Municipal Corporation, set up an integrated MSW (municipal solid waste) processing plant to produce RDF from MSW on 25 acres of land at village Langariwas, about 25 km from Jaipur.
- The plant is designed to process about 500 TPD (tonnes per day) of MSW and generate about 150 TPD of RDF (refuse-derived fuel) in the form of fluff.
- Utilization of RDF has resulted in the conservation of fossil fuels.
- MSW is disposed of in an environment-friendly manner.
- Landfilling requirement for disposing of MSW has reduced by 60%–65%.
- There is an overall reduction in GHG (greenhouse gas) emissions. During the period July 2007–March 2008, the company was able to reduce CO_2 emissions by 7242 MT.

CASE STUDY 9 Orissa Power Generation Corporation Ltd

About the company

The Orissa Power Generation Corporation Ltd, or OPGC, was incorporated on 14 November 1984. The company is a joint venture of Government of Orissa and AES Corporation, USA, a global power generating and distributing company. The main objective of the OPGC is to establish, operate, and maintain large thermal power generating stations independently and in the joint sector.

As its maiden venture, the company has set up a thermal power plant, namely, Ib Thermal Power Station, or ITPS, with 2 × 210 MW capacity in the Ib valley area of Jharsuguda district in the state of Orissa. The power plant is in close proximity to the coal mines as well as to the Hirakud reservoir. This gives the company the distinct advantage of low-cost generation.

The entire generation from these units is committed to GRIDCO (Grid Corporation of Orissa) – the state-owned power transmission – through a long-term Power Purchase Agreement, or PPA. The turnover of the company in 2007/08 was Rs 4770.73 million.

Environmental profile

The company believes in the premise that 'development' cannot be achieved at the cost of the environment. Under this philosophy, the OPGC is committed to produce safe, clean, and reliable power for the country.

The company has adopted a well-defined Environment Policy for minimizing environmental impacts due to the operational activities associated with the generation of thermal power and preserving the natural ecology. It is also in the process of establishing an Environmental Management System as per ISO 14001. The company has established an environment cell for the overall coordination of environmental management and to minimize environmental impact. Besides, an Environmental Monitoring Cell with a well-equipped environmental laboratory is actively involved in the regular monitoring of key environmental parameters, in compliance with the pollution control board norms. The company regularly conducts environmental awareness training programmes to create an environmentally aware workforce.

Air pollution

Air pollutants emitted from the facility include SPM (suspended particulate matter), RSPM (respirable suspended particulate matter), SO_2 (sulphur dioxide), and NO_x (oxides of nitrogen). The facility also faces the problem of odour, noise, and fugitive emissions.

Air pollution control measures

To abate air pollution, state-of-the-art equipment have been installed in the facility. Further, a green belt has been developed in and around the plant and township by planting about

200 000 trees, which act as a major sink to absorb emissions like SPM, RSPM, SO_2, NO_x, fugitive dust and gases, and noise.

SPM and RSPM

- ESPs (electrostatic precipitators) with an efficiency of 99.8% are connected to boiler flue gas path.
- 220-m-tall stacks are provided for the effective dispersion of SPM and RSPM.
- Coal with good calorific value is used as a fuel.
- Various innovative measures are being employed to bring down the SPM level from <100 mg/Nm^3 to <150 mg/Nm^3.

SO_2 and NO_x

- Boilers are provided with low-NO_x burners of tangential coal-firing system to arrest NO_x emissions.
- The low-sulphur coal used by the facility ensures that the SO_2 emissions are within the prescribed standards.

Fugitive emissions

- Cyclone-type dust extractors, dust suppression systems (water sprinklers), and ventilation systems have been installed to control fugitive emissions.
- A dry ash handling system is provided with ash conditioners that prevent the spreading of ash.

Noise control measures

- Regular lubrication and maintenance of equipment ensure their noise-free operations.
- Acoustic hoods have been provided over turbines and generators.
- Double-walled and double-sealed operator cabins have been provided.
- Use of ear plugs has been made mandatory in high noise areas.

Water pollution

The waste water generated from the facility contains BOD (biochemical oxygen demand), COD (chemical oxygen demand), pH, TDS (total dissolved solids), TSS (total suspended solids), oil and grease, and so on.

Water pollution control measures

Effluents like ash water, condenser cooling system blowdown, coal handling plant effluent, and boiler effluent are completely reused in the plant processes. pH of the effluent is brought within the range 5.5–9 by neutralizing the effluent in a neutralization pit. To remove oil from the turbine and transformer yard effluent, it is passed through an oil trap. The treated effluent is used by farmers for irrigation purpose.

Hazardous waste

Hazardous waste generated from the facility includes used lubricants, used batteries, and spent resin. Used lubricants and waste oil are disposed of in the coal yard or sold to authorized recyclers. Used batteries are returned to warehouse, which sells these to battery suppliers or authorized recyclers.

Other waste

Fly ash, bottom ash, coal ash, ferrous and non-ferrous scrap, domestic waste, and biomedical waste are also generated in the facility. Ferrous and non-ferrous scrap is sold to other companies. Domestic waste is treated in a biogas plant. Fly ash is used for making cement, filling low-lying areas, raising the ash dyke in ash pond, and so on.

Case study

Generation of clean and green energy from waste

Introduction

Rapid growth in industries, agriculture, and urbanization, along with population, has led to the generation of huge amount of solid waste. This waste needs to be managed in a holistic manner, so that there are no adverse impacts on the environment as well as human health. An efficient way of disposing of biodegradable waste is treating it in a biogas plant to generate biogas, a green fuel. Biogas is a smoke- and soot-free, clean, non-polluting, and green fuel. It is a mixture of methane, carbon dioxide, hydrogen sulphide, and other gases like nitrogen, hydrogen, and carbon monoxide in traces. But methane forms the major component.

To manage its solid waste from kitchens, canteen, and guest house, the OPGC has established a biogas plant in its premises in 2006, with an installed capacity to process 1 tonne/day of waste, though the facility generates only 0.2 MT of biodegradable waste per day. The generation efficiency of the biogas plant is about 0.08 m^3/kg of kitchen waste.

Principle behind biogas production

Biogas mainly comprises methane. It is produced by anaerobic decomposition in three phases.

- *Hydroysis* Saprophytic bacteria convert complex organic compounds into water-soluble, less complex compounds.
- *Acid formation* Acid-forming bacteria degrade organic compounds into volatile fatty acids and ammonia.
- *Methane formation* Methane-forming bacteria utilize these acids to form methane.

Structure of the biogas plant

The biogas plant at the OPGC is gravity based, so a level difference is maintained in various components of the system. It comprises the following components.

- *Platform* A mixer and premix tanks are located on the platform. Working space is available for segregating and feeding the waste and operating the mixer.

- *Mixer* A 7.5-HP motor is installed on the top of the tank, with a belt pulley to rotate the mixer blade inside the tank.
- *Premix tanks* These tanks have a water supply arrangement for generating slurry.

- *Solar heating system* The solar heating system has a capacity of 250 litres/day. It comprises a storage tank, solar panels, and pipelines. Insulation is provided on pipelines to prevent heat loss.
- *Pre-digester tank* It is an underground cylindrical tank made up of brick masonry, and a concrete block is used as a cover. RCC pipes are provided for the movement of the slurry.
- *Digester* It comprises a main tank, dome, partition wall, an inlet pipe, an outlet pipe, and gas outlet connection.
  - The main tank is an underground structure made up of brick masonry and cement concrete.
  - The dome, or gas holder, is a drum-like structure, fabricated of mild steel. It fits like a cap on the mouth of the digester, where it is submerged in water and rests on the ledge. The drum collects gas, which is produced from the slurry inside the digester. To ensure the flow of stored gas on its own to the point of utilization, the gas is stored inside the gas holder at just above atmospheric pressure. This pressure is achieved by the weight of the biogas-holder dome.
- *Manure pits* These are four underground tanks made of brick masonry.

Process details

Waste collection, segregation, and conditioning

The kitchen waste from the bins is transported to landfill sites, where it is segregated into biodegradable and non-biodegradable waste. Biodegradable waste is carried to the biogas plant for processing. The biodegradable kitchen waste, except paper and raw vegetative matter, is then processed in the 7.5-HP mixer.

Paper is shredded into small pieces for better processing in the plant. It is kept soaked in one of the premix tanks for three to four days. The soaked waste is then passed through the mixer.

Processing of the conditioned waste

The conditioned waste is transferred to the mixer, where it is mixed with water in 1:1 ratio for 10–15 minutes to form slurry. The pH of the slurry is maintained at 6.5–7.5. This slurry then flows to one of the premix tanks in the predigester.

Processing in the predigester

The slurry is subjected to thermophilic bacteria and other micro-organisms in the predigester. The growth of these micro-organisms is ensured by mixing the waste with hot water and maintaining the temperature in the range 55–60 °C. The hot water is supplied by the solar heater. The pH of the slurry drops to 3 in this tank. This is due to the formation of various acids.

Processing in digester

The slurry then enters the digester tank, where it undergoes anaerobic degradation by a consortium of methanogenic bacteria that act on the cellulosic materials of the waste to produce methane. As the gas is generated in the main tank, the dome slowly rises up and reaches a maximum height of 8 feet, holding 35 m^3 of gas. This gas is transported to end-users through a network of pipeline.

Manure

The undigested lignocellulosic and hemicellulosic materials are then passed on to manure pits. These get converted into rich manure, which is then dug out and used for agricultural applications.

Factors affecting the biogas plant

- *Temperature* The optimum temperature for methane-producing bacteria is about 55 °C. The gas production reduces at low temperature and stops at 10 °C.
- *pH* The pH in the predigester is maintained at 6.5–7.5.
- *Retention time* It is temperature dependent. Higher the temperature, faster the bacteria act on the slurry, which then needs replacing.
- *Methanogenic bacteria* Optimum number of these bacteria should be maintained, as the gas production will fall if the number of the bacteria reduces.

Project at a glance

Biodegradable kitchen waste: 200 kg/day
Water (on recirculating basis): 250 litres/day
Electricity (3.7 kW for 30 minutes/day): 1.87 units
Lighting (400 watts for 12 hours): 4.8 units
Biogas from 200 kg waste @ 0.08 m^3/kg of waste: 16 m^3/day
Equivalent LPG (liquefied petroleum gas) produced: 7.75 kg/day
1 m^3 biogas = 0.484 kg of LPG
Digested slurry (manure): 75 kg/day

Cost–benefit analysis

Project cost: Rs 581 000
Labour cost: Rs 73 000/year
Maintenance cost: Rs 30 000/year
Electricity cost for 6.67 units: Rs 7000/year
Water cost: Rs 1000/year
Total: Rs 111 000/year
Cost of biogas produced (@ Rs 58/kg of LPG): Rs 164 000/year
Cost of manure produced (@ Rs 2/kg): Rs 55 000
Total: Rs 219 000/year
Net saving/year: Rs 108 000/year
Payback period: 5 years and 5 months

Benefits

- *Waste management* Biogas plant is an environment-friendly way of managing biodegradable waste generated in the OPGC premises. About 200 kg of domestic waste is now used as a resource.
- *Generation of fuel* A clean and non-polluting fuel, methane, is produced in the biogas plant. This has resulted in the saving of Rs 164 000/year worth of natural resource in the form of LPG.
- *Greenhouse gas mitigation* There are no emissions of CO_2 and methane, potent greenhouse gases, which would have been released due to the natural decay of waste in landfills.
- *Manure* A rich manure is produced, which is used as fertilizer in agriculture.
- *Savings* Net savings per annum due to the production of biogas and manure are to the tune of Rs 108 000.

CASE STUDY AT A GLANCE

- The Orissa Power Generation Corporation Ltd, or OPGC, is a joint venture of Government of Orissa and AES Corporation, USA, a global power generating and distributing company. The main objective of the OPGC is to establish, operate, and maintain large thermal power generating stations independently and in the joint sector.
- The turnover of the company in 2007/08 was Rs 4770.73 million.
- To manage its solid waste from kitchens, canteen, and guest house, the OPGC has established a biogas plant in its premises, with an installed capacity to process 1 tonne/day of biogas, though the facility generates only 0.2 MT (million tonnes) of biodegradable waste.
- Biogas plant is an environment-friendly way of managing biodegradable waste of the OPGC premises. About 200 kg of domestic waste is now used as resource.
- A clean and non-polluting fuel, methane, is produced in the biogas plant. This has resulted in the saving of Rs 164 000/year worth of natural resource in the form of LPG (liquefied petroleum gas).
- There are no emissions of CO_2 (carbon dioxide) and methane, potent greenhouse gases, which would have been released due to the natural decay of waste in landfills.
- Rich manure is produced, which is used as fertilizer in agriculture.
- Net savings per year due to the production of biogas and manure are to the tune of Rs 108 000.

CASE STUDY 10 Hero Honda Motors Ltd

About the company

Established in 1984, Hero Honda Motors Ltd is a joint venture between Honda Motor Co., Japan, and Hero Honda Group, India. It is the leading two-wheeler manufacturing company in the world. It is the first company in India to introduce a new generation two-wheeler that has set industry benchmarks for fuel economy and low emissions.

The company has manufacturing facilities at Dharuhera and Gurgaon in Haryana. To cater to the growing market demand, the company is in the process of setting up another manufacturing unit in Haridwar, Uttarakhand. Its annual turnover is around Rs 100 billion.

Environmental profile

Being a leader in the automotive industry, the company is fully aware of its responsibility towards maintaining a cleaner and pollution-free environment. It has an Environment Policy at both facility and corporate levels. The policy lays down broad guidelines for the company towards achieving excellence in its environmental performance, as mentioned below.

- Integrate environmental attributes and cleaner production in all business processes and practices, with specific consideration to substitution of hazardous chemicals.
- Continue product innovations to improve environmental compatibility.
- Comply with all applicable environmental legislation and control all environmental discharges.
- Institutionalize resource conservation
- Enhance environmental awareness in all employees, dealers, and vendors.

The company has a certified Environmental Management System since 1999, under which it has implemented a number of environmental initiatives. There is also an environment laboratory in the plant, which monitors pollution control facilities. The company has also undertaken various energy conservation initiatives like using energy-efficient lighting and motors, automation in electric circuit, and so on. It has also phased out ozone depleting substances from its operations.

The company has been certified ISO: 14001 for its Environmental Management System, OHSAS 18001 for its Occupational Health and Safety Management Systems, and ISO 9001 for its Quality Management Systems.

Air pollution and its control measures

The air pollutants emitted from the facility include SPM (suspended particulate matter), RSPM (respirable suspended particulate matter), SO_2 (sulphur dioxide), NO_x (oxides of nitrogen), and fugitive dust. The power generator sets and boilers have been provided with chimneys of appropriate height to ensure efficient dispersion of pollutants. Bag filters and scrubbers have also been installed to control emissions.

Water pollution and its control measures

The main water pollutants discharged from the facility are BOD (biochemical oxygen demand), COD (chemical oxygen demand), pH, total suspended and dissolved solids, heavy metals, and oil and grease.

State-of-the art facilities have been set up for treating industrial and domestic effluents. Industrial treatment plants have the provision for the automatic dosing of chemicals and online monitoring of flow and other parameters. The plants are supported with in-house environmental laboratories equipped with latest instruments. The treated water is used for horticultural purposes.

Hazardous waste

Hazardous waste generated from the facility includes paint sludge, phosphate sludge, acid bath residue, waste solvents, paint-bearing waste, oil-contaminated waste, waste oil, heavy metal bearing waste, incinerator ash, and waste batteries, among others.

Paint and phosphate sludge, incinerator ash, and acid bath residue are disposed of in secured landfills. Waste solvents, oil-contaminated waste, and paint-bearing waste are incinerated under controlled conditions. Waste oil and batteries are sold to authorized oil recyclers. Heavy metal bearing waste is stabilized in an ETP (effluent treatment plant), and the sludge generated is disposed of in secured landfills.

The company has devised many innovative methods to reduce the generation of hazardous waste, like substitution of lime with caustic soda, which is used as a dosing material in the ETP, to reduce the generation of waste water sludge, recovery of solvent from the waste-cleaning solvent in the painting process, and so on.

Other waste

Other industrial waste items generated from the facility include waste electronic material, metal and non-metal scrap, waste aerosol spray cans, rubber components, waste resin, paper waste, and packaging waste. All these are sold to authorized recyclers.

Case study
Recycling of the paint sludge

Background

Painting is a part of the manufacturing process in an automobile industry. In the spray painting process, the off-spray paint, which does not deposit on the components, gets wasted and is collected in the form of paint sludge, which is a hazardous waste. As per the hazardous waste rule, the paint sludge is required to be incinerated under control conditions. The residual ash thus generated is also required to be disposed of in secured landfills. The process of incineration releases harmful toxic gases, causing air pollution and threatening human health. The process of landfilling is also very expensive.

The company, on an average, consumes 760 litres of paint per 1000 units of vehicles produced and generates approximately 597 kg of wet paint sludge (40% moisture). The volume of paint sludge generated has increased over the years due to increased production capacity. Management of this waste was increasingly becoming difficult because of the absence of a common treatment, storage, and disposal facility in the state. Costs of incineration and disposal were also becoming prohibitive. To address these issues, the company, along with the M/s Maharani Paints India Pvt. Ltd, after a lot of deliberations and trials, implemented a novel method for recovering paint from the waste paint sludge. The company has obtained special permission from the Haryana State Pollution Control Board to recycle up to 25 tonnes of paint sludge per month. The commercial supply of recycled paint started in October 2006.

Process description

The paint sludge contains paint pigment, one of the basic paint ingredients. This pigment is recovered to make a paint primer, which has the quality of virgin material. A batch of 500 kg of paint sludge and 650 kg of additives is processed to produce 732 litres of anti-corrosive paint primer, generating 2%–3% of paint sludge (Figure 1). The recycled primer is used in the painting process, and the sludge is incinerated.

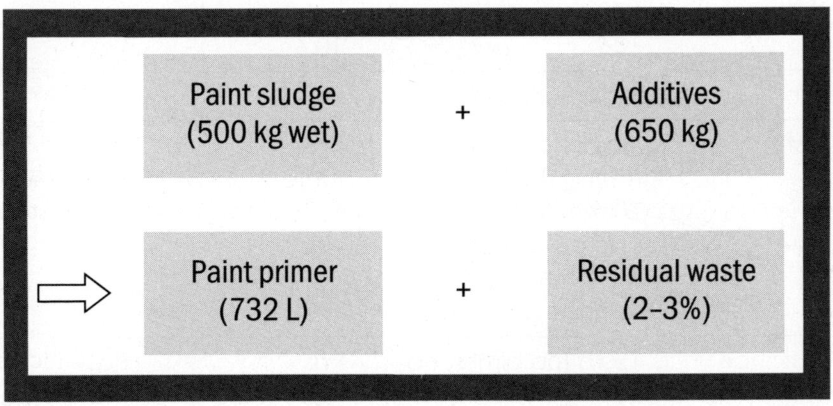

Figure 1
Production of the paint sludge

The paint sludge is washed for removing water-soluble impurities. It is then dried in two stages to remove the moisture content. Following this, the sludge is soaked in chemicals, which soften it. It is then subjected to dispersion stirring and blending for homogenization with additives, filtration for the removal of large particles, milling and grinding to bring it in the form of paint, and final correction to adjust viscosity and other parameters. The recycled primer thus obtained is tested for quality and packed in steel containers. Figure 2 gives the process detail.

Table 1 gives a comparison of recycled anti-corrosive primer with virgin plant primer.

Table 1 A comparative account of recycled anti-corrosive primer and virgin plant primer

Property	*Specification*	*Virgin paint primer*	*Recycled anti-corrosive primer*
Viscosity (sec) at 300 °C	65±5	64	63
% N.V.M.	50±2	50.1	49.18
Gloss at 600	Minimum 20	28	30
Application viscosity (sec)	15–17 (B4 Ford Cup @30 °C)	16	16
Adhesion	To pass 100/100	Pass	Pass
Impact resistance	500 g/20 cm	Pass	Pass
Flexibility test	No cracks or flaking	Pass	Pass
Gasoline test	To pass 7 hours	Pass more than 8 hours	Pass more than 8 hours
Water resistance WT/10 LTR (kg)	120 h	Pass 120 h	Pass 120 h
11±0.5	10.9	10.6	
Thinner intake	Maximum 80% v/v	74%	75%
Salt spray test	No blister, no peel, rusting shell within 20 mm from either side of scratched line after 120 h of testing	Pass More than 120 h	Pass More than 120 h
Humidity resistance	120 h (no blister, no flaking wrikless, rust, excessive change in colour or softening of coating)	Pass More than 120 h	Pass More than 120 h
Resistance of thinner paint of application (K OHMS)	500–100 (resistivity meter)	694	725

Cost–benefit analysis

Cost of standard anti-corrosive primer: Rs 117/litre

Cost of recycled anti-corrosive primer: Rs 108/litre

Disposal cost of the paint sludge: Rs 13.66 per litre of primer

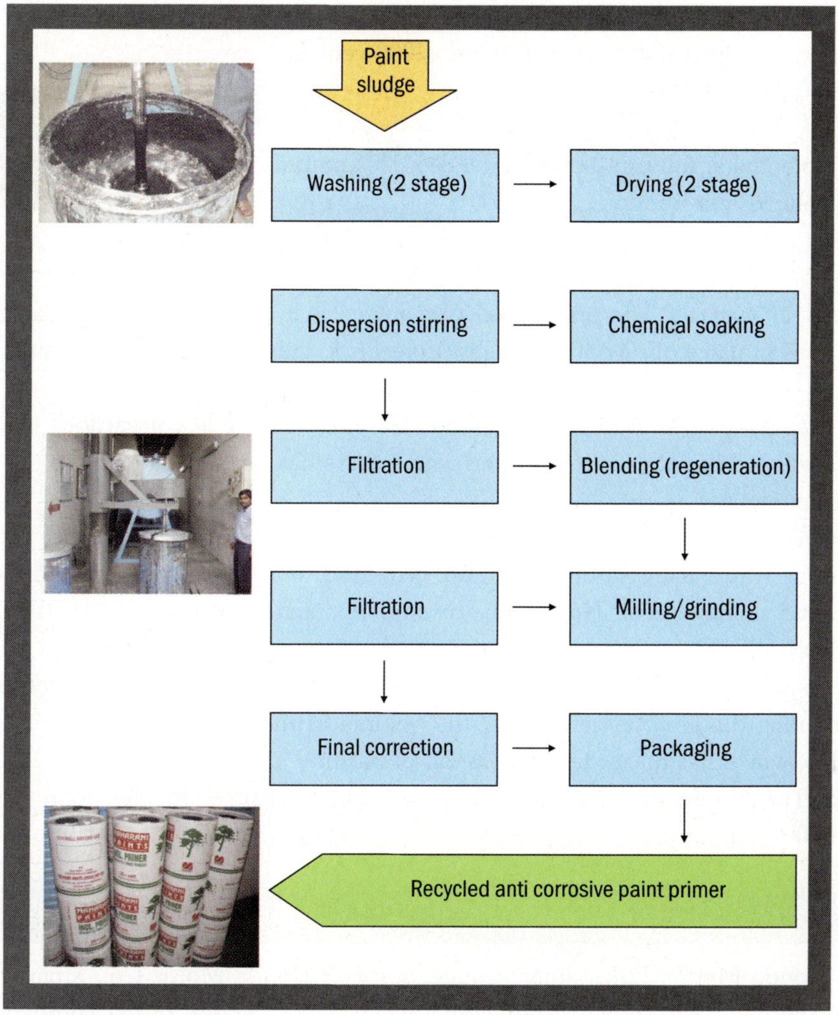

Figure 2
Paint sludge recycling process

[Quantity of paint sludge recycled for producing 1 litre of primer: 0.683 kg
Treatment and disposal cost of the paint sludge: Rs 20 per kg of sludge]
Net saving by using recycled primer: Rs 22.66/litre

Saving potential

For use in frame body only

Consumption of anti-corrosive primer: 68 litres/1000 vehicles
Saving: Rs 1541/1000 vehicles

For use in fuel tank only

Consumption of anti-corrosive primer: 62 litres/1000 vehicles
Saving: Rs 1405/1000 vehicles

For use in frame body and fuel tank
Consumption of anti-corrosive primer: 130 litres/1000 vehicles
Saving: Rs 2946/1000 vehicles

Total saving potential at full production: Rs 8 838 000/annum
(@ 3 million vehicles per annum)

Benefits

- *Conservation of resources* The project recovers anti-corrosive primer from the paint sludge, which is used in place of raw primer. This project has, thus, led to the conservation of raw material.
- *Reduced quantity of hazardous waste generated* The paint sludge is a hazardous waste. Due to the recovery of useful primer from this waste, there has been a reduction in the volume of sludge generated.
- *Waste management* Due to the increase in the production capacity, the generation of hazardous waste paint sludge had also increased, posing a problem for its efficient handling and management. Now, with the implementation of this project, there is less waste generated, which is managed easily.
- *Reduced air pollution* Incineration of huge quantity of paint waste used to release toxic gases, polluting the environment. Now, the volume of the waste generated has reduced remarkably, which has meant less generation of harmful air pollutants.
- *Economics* The implementation of this project has resulted in the total saving of Rs 8838000 per annum.

CASE STUDY AT A GLANCE

- Hero Honda Motors Ltd, a joint venture between Honda Motor Co., Japan, and Hero Honda Group, India, is the leading two-wheeler manufacturing company in the world.
- The company generates hazardous waste in the form of paint sludge.
- Management of this waste was increasingly becoming difficult because of the absence of a common treatment, storage, and disposal facility in the state. Costs of incineration and disposal were also becoming prohibitive.
- To address these issues, the company, along with the M/s Maharani Paints India Pvt. Ltd, has implemented a novel method of recovering paint (primer) from the waste paint sludge.
- This project has thus led to the conservation of raw material.
- Due to the recovery of useful primer from this waste, there has been a reduction in the volume of sludge generated, which has meant easy and low-cost management of waste and less generation of harmful air pollutants.
- The implementation of this project has resulted in the total saving of Rs 8 838 000 per annum.

CASE STUDY 11 Vadodara Manufacturing Division

About the company

The Indian Petrochemicals Corporation Ltd, or the RIL-VMD (Reliance Industries Ltd-Vadodara Manufacturing Division), amalgamated as Reliance Industries Ltd, or RIL, is a leading enterprise, manufacturing a range of petrochemical products such as polymers, synthetic fibres, surfactants, industrial chemicals, catalysts, and absorbents since 1972, with a multi-location manufacturing activities at Baroda, Nagothane, and Gandhar.

RIL-VMD is the oldest complex that has 16 plants and manufactures 39 products. The complex is located in an area of 774 acres and is operating since 1973. The annual turnover of the company in 2007/08 was Rs 1392.69 billion.

Environmental profile

RIL-VMD has an established EMS (Environmental Management System) of international standard, based on ISO 14001. Accreditation was done by M/s Det Norske Veritas, the Netherlands.

Environmental monitoring is an important part of the EMS. Essentially, it operates on three-tier monitoring.

- *First tier* Internal monitoring carried out by a well-equipped laboratory established by RIL-VMD. Regular audits are carried out by well-trained internal auditors for ISO 14001, and a unique green card rating system has been developed for Baroda complex.
- *Second tier* Monitoring is carried out by external agencies, such as the GPCB (Gujarat Pollution Control Board)-approved auditors.
- *Third tier* Monitoring by the GPCB and the CPCB (Central Pollution Control Board).

Air pollution

The main air pollutants generated from the facility are SPM (suspended particulate matter), SO_2 (sulphur dioxide), NO_x (oxides of nitrogen), and fugitive emissions.

Air pollution control measures

RIL-VMD has installed state-of-the-art air pollution control equipment in various plants. Stack emissions from furnaces, heaters, and boilers are maintained well below the specified standards by using low-sulphur natural gas supplied by GAIL Ltd, internally generated fuel gas, and LSHS (low sulphur high stock) fuel oil. Efforts are on to minimize the usage of LSHS, subject to the availability of natural gas. Apart from this, stacks of adequate height are provided to ensure proper dispersion of stack emissions.

Emissions from process equipment, vents, and valves are connected to the flare system. Four flare stacks, each of 100 m height, are installed with steam atomization system to ensure complete combustion of the gases emitted during normal run as well as during upsets. The plants are also equipped with scrubbers, dust elimination system, and mist eliminator.

Water pollution

The main water pollutants discharged from the facility are BOD (biochemical oxygen demand), COD (chemical oxygen demand), pH, oil and grease, and TSS (total suspended solids).

Water pollution control measures

RIL-VMD has state-of-the-art waste water treatment facilities, which include the following.

- ISBL (inside battery limit) treatment facilities for the pre-treatment of waste from different process plants to remove specific pollutants have been set up.
- Second stage treatment at the CWWTP (central waste water treatment plant) comprises common sump, screw pumps, screens and grit chamber, surge ponds, API (American Petroleum Institute) oil separator, homogenization tank, clariflocculator, aeration tank, final clarifier, sludge recycling sump, sludge thickener, sludge drying beds, final effluent pumps, pump house, and so on (Figure 1).

Figure 1 The RIL-VMD effluent treatment plant

- The waste water from the ACN (acrylonitrile) plant is segregated and treated separately by processes developed and patented in-house to meet the stringent requirements. A new ACN waste water bio-treatment plant (300 m^3/day) has been commissioned to replace the old biological treatment plant.
- The treated effluent, with all parameters within stipulated standards, is discharged into the Gulf of Cambay, near the estuary of the River Mahi, through a first-of-its-kind unique and efficient effluent channel project.
- A tertiary plant (2 million gallons per day capacity) for treating effluent for reuse in cooling water make-up and horticultural activities is in place.

Hazardous waste

The major categories of hazardous waste generated are furnace/reactor residue and debris, tarry residue, oily sludge emulsion, organic residue, spent catalysts and molecular sieves, oil emulsion, used/spent oil, toxic-metal-containing residue, and so on.

Hazardous waste management

RIL-VMD has an efficient hazardous waste management system in place.

- A state-of-the-art double-lined secure landfill has been designed, with a capacity of 12 800 MT (million tonnes). It comprises side and bottom layers, and a leachate collection and detection system.
- A multipurpose incinerator to handle 2000 MTPA (million tonnes per annum) of solid/ semi-solid/liquid waste is in operation since 1998. A dedicated incinerator with 500 MTPA capacity is also functional since 1983. The incinerators are provided with air control equipment and a waste heat recovery steam generator.
- Manifest system is followed for keeping track of the waste disposed.

Case study

Reduction of hazardous waste generation at RIL-VMD

Objective

Hazardous waste generation in the VMD complex was a problem since 1989. During 2002/03, hazardous waste generation at RIL-VMD was about 5673 MTPA. As a result, the company embarked on a mission to reduce the generation of hazardous waste at its site by employing various modern technologies like Six Sigma Methodology and cleaner production. The facility aimed at not only reducing the total generation of hazardous waste, but also halving the specific generation of hazardous waste from its value in 2002/03.

Hazardous waste reduction by cleaner technology

Reduction in the generation of slop oil from linear alkyl benzene plant

In the UOP processing route for the production of LAB (linear alkyl benzene), a precursor for LAS (linear alkyl benzene sulphonate), linear paraffins, a mixture of nC_{10}–nC_{13}, are catalytically dehydrogenated to corresponding linear olefins in a UOP Pacol process unit. The effluent from this unit is then alkylated with benzene in a UOP detergent alkylation unit in the presence of hydrogen fluoride catalyst to produce LAB.

Small amounts of di-olefins are also produced in the Pacol reactor. The di-olefins form undesirable by-products, tar or slop oil, when alkylated, decreasing the yield of LAB and degrading LAB quality. Processing the Pacol separator liquid in the DeFine unit selectively converts the slop oil to the desired mono-olefine/olefin. The DeFine unit primarily consists of a reaction vessel containing the proprietary DeFine catalyst. In the typical UOP-designated LAB complexes, the DeFine reactor is placed between the Pacol separator and the Pacol stripper. The DeFine process significantly improves the overall efficiency and profitability of an LAB complex operation. Figure 2 shows the benzene butadiene hydrogenation plant.

Figure 2
The RIL-VMD benzene butadiene hydrogenation plant

Project description

DeFine is the selective hydrogenation of di-olefins to mono-olefins for optimal LAB yield and quality. In the Pacol reactor, along with mono-olefins, di-olefins are also produced. When alkylated, these di-olefins form low-value by-products like HAB (heavy alkylate benzene) and polymers and also degrade the quality of LAB. The DeFine unit converts most of these di-olefins back to mono-olefins and paraffins. The DeFine reactions are carried out in liquid phase, over a fixed bed of catalyst and under controlled conditions of temperature. The process flow diagram is shown in Figure 3.

Cost–benefit analysis

Total expenditure incurred in implementing this project was Rs 76.5 million. Benefit accrued is Rs 105 million per annum. So, the payback period is approximately six months.

Benefits

- There is a reduction in the formation of low-value by-products. Slop oil generation reduced by about 96% after the implementation of the project (Figure 4).
- There is an increase in the production of LAB.
- There is an improvement in the quality of LAB.
- There is a reduction in the specific consumption of raw materials.
- There is an extension in the run length of Pacol catalyst.

Reduction in the generation of slop oil from the naptha cracker plant

Process description

In the naphtha cracker plant (Figure 5), the third stage was cooled (outlet temperature 55–57 °C) in a discharge cooler, after which the hydrocarbon condensate was knocked off in a knock-out vessel. This stream was then routed to the caustic tower for the removal of acid gases. The heavy aromatic wash oil (Remax) and heavy hydrocarbons used to condense in the

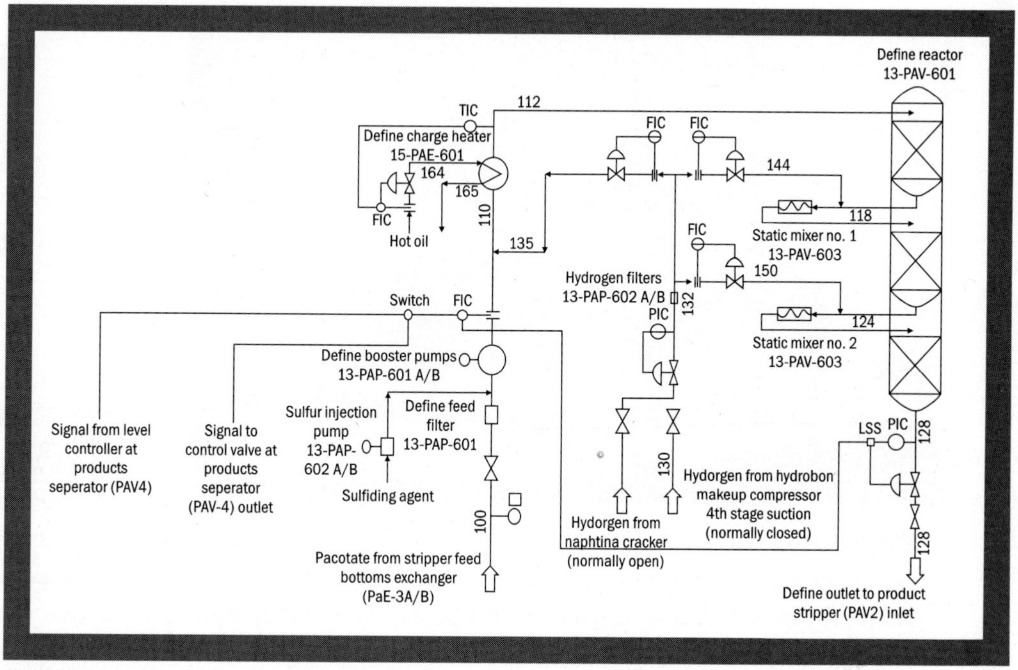

Figure 3 Process flow diagram showing the conversion of di-olefins to mono-olefins

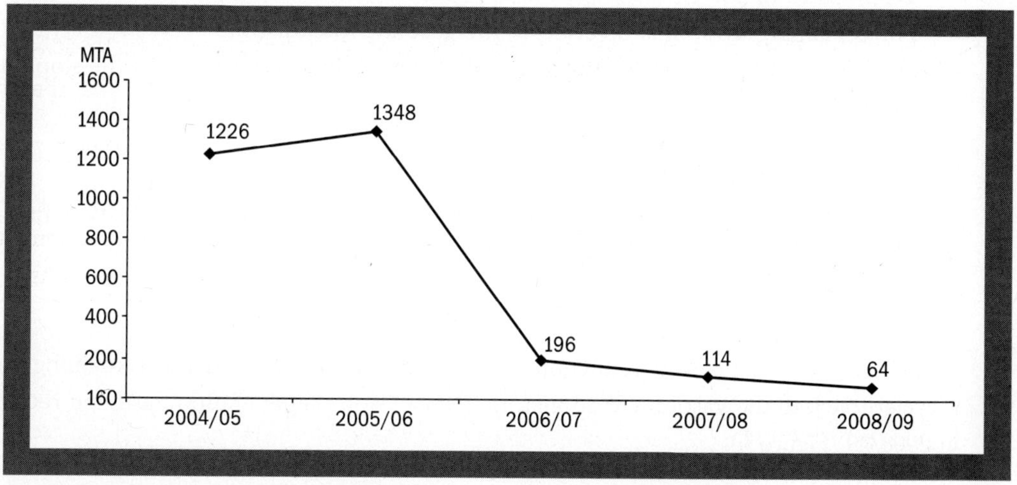

Figure 4 Reduction in slop oil generation in the linear alkyl benzene plant

caustic tower. These condensed hydrocarbons from the bottom of the caustic tower, along with the spent caustic, were diverted to the de-oiling drum. Here, the polymeric oil was extracted by using fresh naphtha in the de-oiling drum.

Sources of slop oil

- Hydrocarbon condensed in the caustic tower.
- Condensing Remax from compressor in the caustic tower.
- Fresh naphtha used for extracting polymeric oil.

Figure 5
The RIL-VMD naphtha cracker plant

Technological changes

- Discharge cooler replaced with high-capacity exchanger, and the outlet temperature was decreased to 47 °C, due to which major part of condensation of hydrocarbon and Remax takes place in the knock-out vessel 12–204.
- Caustic tower temperature is maintained at the overhead temperature of 49 °C and bottom temperature of 56 °C.
- The remaining slop oil is diverted to the gasoline fractionator.
- Boiler feed water is used for the removal of soluble impurities to make the slop oil suitable for gasoline pool.

Cost–benefit analysis

Total expenditure incurred on implementing this project was Rs 100 000. Benefit realized is Rs 4.5 million per annum. So, the payback period is 10 days.

Benefit

There has been a marked reduction in slop oil generation, and the remaining slop oil is recycled back to the gasoline fractionators. Hence, slop oil generation has been reduced to nil in 2008/09 (Figure 6).

Hazardous waste reduction by Six Sigma Methodology

Reduction in the generation of EDC bottoms from the vinyl chloride plant

Project description

Reduction of hazardous waste (EDC [ethylene dichloride] bottom) from 0.03 MT/MT of VCM (vinyl chloride monomer) to 0.02 MT/MT of VCM. EDC bottom generation takes place in the cracker and HTC (high-temperature chlorination) area of the vinyl chloride plant. EDC bottom is finally separated in tar still (cracker and HTC) and transferred to a storage tank. By adopting Six Sigma tool DMAIC, the generation of hazardous waste EDC bottom was reduced from 50% to 17%–22%. Figure 7 shows the reduction in the generation of EDC bottom during the period 2004/05–2007/08.

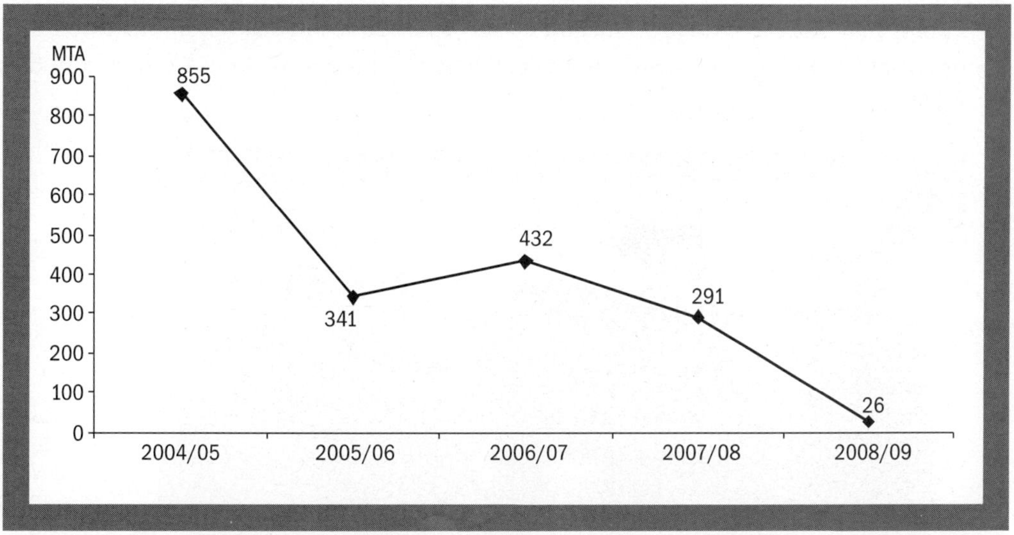

Figure 6 Slop oil generation in Naptha Cracker Plant

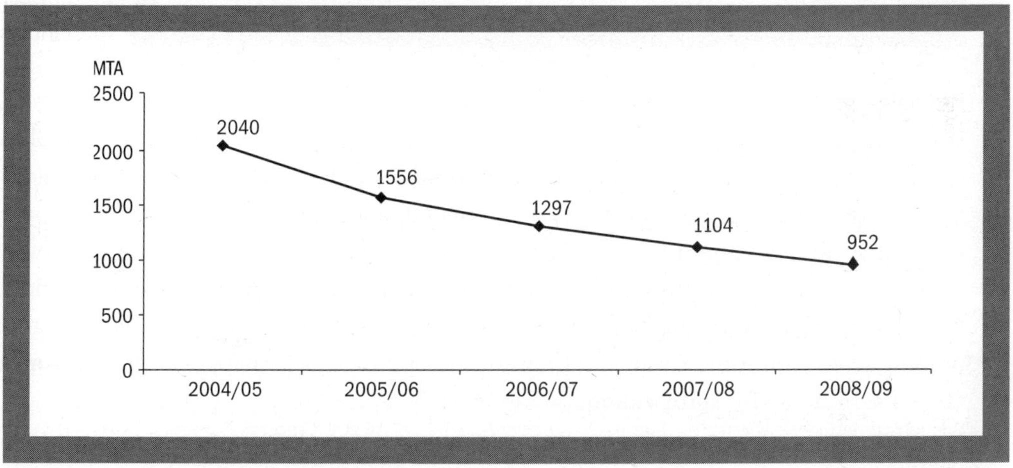

Figure 7 Reduction in the generation of EDC bottoms

Cost–benefit analysis

The total expenditure incurred was nil. The benefit accrued is Rs 4.4 million.

Benefit

There is a 43% reduction in the generation of the hazardous waste EDC bottom.

Reduction in the generation of reactor waste from the polybutadiene rubber plant

Project description

Increasing the size of suction spool piece

Five reactors having individual Sierbath bottom pumps are used for transferring cement slurry from one reactor to the other reactor. Due to their current dimensions, the pumps do not get sufficient suction, and the suction spool piece gets fouled with cement. As the speed of the pumps increases with the level in the reactor, there is a possibility of the dry run of pumps.

Pumps may not have flooded suction all the time. To avoid starvation of suction to pumps, a proposal to modify the suction spool piece just like hopper was put forth. Figure 8 shows the polybutadiene rubber plant

Figure 8
The RIL-VMD polybutadiene rubber plant

Reactor internal surface buffing

Due to the unavoidable side polymerization reaction in the poly reactors, there is always some amount of continuous gel formation or by-product polymer formation. This gel/by-product has a tendency to settle on walls, get accumulated, and result in big lump formation. This can choke and obstruct liquid flow in the pump cavity, and suction and discharge pipes, and can result in further loss of pumping capacity. After a lifespan of one to one and a half months, this reactor, along with the bottom pump, is required to be changed. This change is always accompanied with some waste and offgrade generation. Such changeover in reactors generally takes 12–15 hours for stabilization.

With better surface finish like mirror finishing of the surface in contact with the liquid in the reactor, the adhesion of insoluble gel to the wall was significantly reduced, which, in turn, led to reduced lump formation and better life cycle of reactors. This also reduced waste and offgrade product formation and improved productivity.

Cost–benefit analysis

For project 'increasing the size of suction spool piece', the total expenditure incurred was Rs 100 000 and for the project 'reactor internal surface buffing', the total expenditure incurred was Rs 250 000. The benefit accrued is Rs 2.4 million per annum. So, the payback period is two months.

Benefits

Implementation of above projects improved pump performance and productivity, and reduced poly shutdown. The reactor changeover frequency due to fouling and cement header replacement frequency also reduced. There was a marked reduction in reactor waste generation and land pollution, and there is a less load on incinerators.

Reduction in the generation of HBB from the polybutadiene rubber plant

Project description

After stripping of cement, the unconverted mix hydrocarbons and impurities go to the three-packed tower unit for hydrocarbon recovery, where benzene and butane are separated and recycled to the feed blending system.

In the final recovery in T-420, benzene is recovered from the top and the remaining impurities are withdrawn as HBB (high boiler bottom)-1 by-product from the bottom. It is essential to remove this by-product from the system continuously, as it will affect the polymerization reaction. HBB is a mixture of benzene, dinmer, timer, and some antioxidants.

To reduce the generation of HBB, instead of continuous withdrawal of HBB from T-420 bottom to Tk-606, withdrawal was done based on T-420 bottom temperature. Normally, bottoms are removed when the benzene concentration is approximately 74% by weight and bottom temperature is 115 °C. When more steam is given to the tower, more benzene gets evaporated. The temperature remains constant at 120 °C and does not increase beyond this. At this point, benzene is withdrawn. With this minor change, there is an increase in the rate of benzene recovery, enhancement in tower performance, and reduction in waste generation.

Cost–benefit analysis

The total expenditure incurred for implementing this project was nil. The total benefit accrued is Rs 2.4 million per annum.

Benefit

With batch-wise withdrawal, instead of continuous withdrawal, HBB 1 waste generation reduced from an average of 2.5 kg/MT to 1.1 kg/MT, resulting is 56% reduction.

Overall benefits

- Quantity of hazardous waste is reduced significantly (Figure 9).
- There is an increase in production/manufacturing efficiency.
- There is a decreased exposure of the workforce to hazardous waste.
- The usage of incinerators decreases because of which there are less emissions.

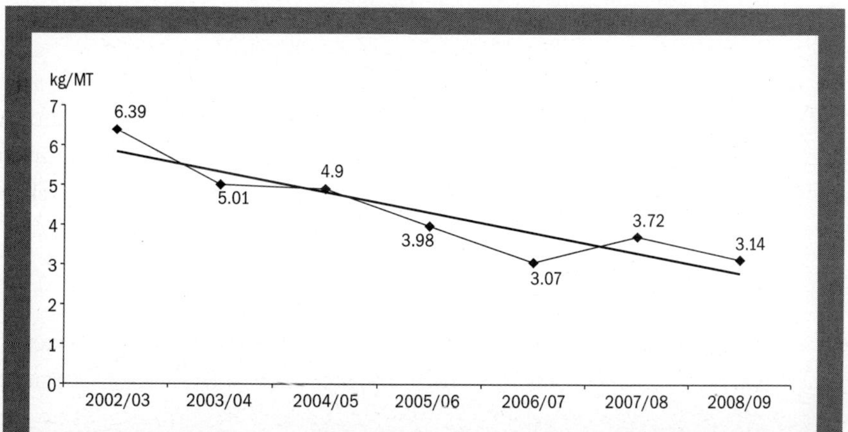

Figure 9
Reduction in the hazardous waste generation

CASE STUDY AT A GLANCE

- The RIL-VMD, amalgamated as Reliance Industries Ltd, or RIL, is a leading enterprise, manufacturing a range of petrochemical products such as polymers, synthetic fibres, surfactants, industrial chemicals, catalysts, absorbents, and so on.
- The annual turnover of the company in 2007/08 was Rs 1392.69 billion.
- The company has embarked on a mission to reduce the generation of hazardous waste at its site by employing various modern technologies like Six Sigma Methodology and cleaner production.
- There has been 50% reduction in ethylene dichloride bottoms generation from the VCM (vinyl chloride monomer) plant, 50% reduction of slop oil generation from the LAB (linear alkyl benzene) plant, 100% elimination of slop oil from the NCP (naphtha cracker plant), and about 60% reduction in waste generation from the PBR (polybutadiene rubber) 1 plant.

CASE STUDY 12

Century Pulp and Paper

Figure 1
Century Pulp and Paper

About the company

Century Pulp and Paper, a unit of the top Indian corporate conglomerate B K Birla Group of Companies, is one of the leading integrated pulp and paper plants and was established in 1984 at Lalkua, a township near the Himalayas, in the state of Uttarakhand (Figure 1). After undertaking extensive expansion and modernization works, the company presently operates four fibre lines, as mentioned below.

- Rayon grade/paper grade pulp: 31 320 tonnes per annum capacity
- Writing and printing paper (wood): 37 250 tonnes per annum capacity
- Writing and printing paper (bagasse): 84 600 tonnes per annum capacity
- Writing printing paper (recycled): 75 960 tonnes per annum capacity

The company has established a strong brand image in the domestic and overseas market and is known for the excellent quality of its products and services. It caters to diversified sectors that manufacture note books and diaries, continuous stationary, copier, envelops, offset printing, security papers, industrial papers, viscose filament yarn, and cellophane papers. The turnover of the company was Rs 8457.2 million in 2007/08.

Environmental profile

The company has been awarded EMS: ISO-14001 Certificate for adopting cleaner production, utilizing resources efficiently, and using waste minimization techniques. Under the auspices of the USAID (United States Agency for International Development) environment programme, Century Pulp and Paper was selected as a model unit by the Confederation of Indian Industry-International Resource Group, or CII-IRG, for implementing EMS: ISO-14001 Certification.

Environment policy

There is an Environment Policy in place for implementing and improving the company's Environmental Management System. The main objective of the policy is to recycle resources, minimize waste, ensure compliance with applicable regulations, continually improve, and create environmental awareness amongst employees.

Awards/accolades

- The company has been awarded the golden status by the DGFT (Director General of Foreign Trade), Government of India, as the Golden Star trading House at the corporate level.
- The company was selected as a model unit for waste minimization by the CPCB (Central Pollution Control Board)/NPC (National Productivity Council).
- The company was adjudged as an efficient unit in energy conservation by the CII in 2004/05.
- In 2004/05, the company won the first prize for energy conservation instituted by the Indian paper Makers Association as well as the national award for energy conservation in the pulp and paper sector.
- The company was given the National Award on Excellence in Energy Management 2007 and 2008 by the CII–Sohrabji Godrej GBC.

Air pollution

The main air pollutants emitted are SPM (suspended particulate matter), RSPM (respirable suspended particulate matter), SO_2 (sulphur dioxide), NO_x (oxides of nitrogen), and fugitive dust and gases. Unpleasant odour is also a problem.

Air pollution control measures

Air pollutants

Electrostatic precipitators, the sophisticated control devices, are attached with stacks to curb air pollutants. These also recover useful chemicals like sodium sulphate. Bag filters are also used. All electrostatic precipitators and bag filters run at optimum efficiency.

Odour

Odour in bagasse-based plant is controlled by using a continuous digester with cold blow system. A new design blow heat recovery system has been installed in the wood-based plant, and a double jet condenser system has been installed in the rayon grade plant. Further, prehydrolysate liquor is burnt in recovery boilers instead of treating it in lagoons.

Fugitive emissions

Fugitive emissions are controlled by operating the plant effectively and efficiently. Using water sprinklers and scrubbing systems and periodic maintenance of machineries have also been found effective in controlling fugitive emissions.

Noise

For controlling noise, the company has adopted engineering control methods in terms of reduction in velocity, pressure, and air turbulence. Personnel protective equipment such as ear plugs are also provided to employees.

Water pollution

The main water pollutants released from the facility are BOD (biochemical oxygen demand), COD (chemical oxygen demand), oil and grease, TSS (total suspended solids), and pH.

Water pollution control measures

The company has installed an ETP (effluent treatment plant), which comprises primary clarification, anaerobic lagooning, biological treatment through aeration and return sludge recycle, and finally, secondary clarification. The effluent from the bagasse-based paper machine, wood-based machine, and deinking-based machine is recycled to the pulp mill through clarifiers. The overflow is taken to the biological reactor to remove the sludge.

The effluent from the rayon grade plant is divided into three streams: prehydrolysate liquor, coloured effluent, and colourless effluent. The prehydrolysate liquor is neutralized with caustic and burnt directly in the chemical recovery boilers after mixing with black liquor. This has resulted in a significant reduction in the pollution load going to the ETP. Also, the burning generates steam, which can be used in the plant. The coloured effluent is taken to the primary clarifier. The overflow goes to the lagoon. The colourless effluent is directly taken to the biological reactor, which is equipped with aerators. After aeration, the effluent goes to the secondary clarifier. The sludge is recycled to the biological reactor. Excess sludge from the biological reactor is pumped to the sludge handling system.

Hazardous waste

Used lubricant oil is the only hazardous waste generated in the company. The used oil is stored in the store yard. A collecting pit is also used for the same purpose. This oil is sold to authorized recyclers. The company has also undertaken research and development for using used oil in the company's process system. The used oil is being used in rotary lime kiln in the chemical recovery plant on a trial basis.

Other waste

Other waste generally comprises industrial waste, including wood dust, deinking plant waste sludge, ETP primary sludge, bagasse pith, lime sludge, and fly ash. Lime sludge is recalcined in a rotary lime kiln to make lime powder. Lime sludge with high silica content is disposed of through contractors. Coal ash and wood dust are also disposed of to outside parties. The ETP sludge from primary clarifiers is recovered and sent to outside parties for use as raw material. The pith generated from the bagasse plant during dry depithing is burnt in the boiler. The wet pith generated is dewatered using a screw press, which is then utilized as a fuel in the boiler. The deinking plant waste sludge is disposed of to board mills.

Case study

Installation of a dewatering screw press for dewatering the wet pith to enable its use as a fuel in boilers

Background

In Century Pulp and Paper, bagasse, which is generated in sugar mills, is used as a raw material in bagasse-based pulp manufacturing process. The bagasse contains pith, which is separated from the bagasse in a depither in two subsequent phases–dry depithing and wet depithing. The depithed bagasse obtained after dry depithing still contains some amount of pith, which is separated further in the wet depithing process. Earlier, the wet pith with 90% moisture content used to be disposed of on low-lying land, which used to emit methane, a greenhouse gas, during its decay.

The wet pith obtained from the wet depithing process has substantial calorific value. The wet pith is subjected to dewatering using a specially designed screw press customized for the process to reduce the moisture content of the pith from 90% to 50% (Figure 2). The pith so obtained is directly co-fired with coal as a fuel in a new specially designed pith-cum-coal-fired boiler for producing steam, which, in turn, is used to generate power. This leads to the substitution of an equivalent amount of greenhouse gas emitting coal as fuel in the co-generation unit.

Figure 2
Wet pith dewatering screw press

Objective of the project

- Avoidance of methane emission from the decay of organic mass, that is, pith.
- Utilization of the calorific value content of the pith for generating steam in a pith-cum-coal-fired boiler, leading to the substitution of an equivalent amount of greenhouse gas emitting coal as fuel in the co-generation unit.

Wet pith dewatering system

The wet pith dewatering system consists of a Fasster make dewatering screw press. The screw press is designed to compact fibrous sludge, and the unit produces a dry product suitable for mechanical handling. It is a horizontal dewatering extrusion assembly. It comprises the following units.

- A screen in three parts of equal length, covered by hoods to confine any water ejected by the sludge compaction process. The filtrate is recovered in a sump located below the screen. The collection pan has a connection flange for filtrate to reach the plant liquid disposal point.
- An extruder (Archimedian Screw) mounted on roller bearing and a self-aligning bearing transferring, dewatering, and extruding the sludge.
- An outlet flap valve with flexible jacks operated by a pressure distribution network or a compressed air generation system.
- The jacket valve assembly to apply the necessary pressure for sludge compaction.
- The jacket pressure maintained by an adjustable pressure relief cum regulating valve.

Coal-cum-dry-pith-fired boiler

The dry pith obtained from the wet pith dewatering system cannot be burnt alone in the boiler for generating steam. It always requires a supporting fuel like coal. The technology for burning pith and coal in the boiler was not available. So, a tall boiler with controlled combustion zone, designed to facilitate the burning of various types of fuels, was procured. The new boiler installed generates steam at 60 tonnes per hour at a pressure of 62 kg/cm^2 and temperature of 485 °C (Figure 3).

Figure 3
Pith-cum-coal-fired boiler

Turbogenerator set

The 14-MW additional power requirement of the plant is met through power generated in a newly installed 16-MW double extraction cum condensing turbogenerator set with one controlled and one uncontrolled extraction (Figure 4).

Figure 4 Turbogenerator set

Project at a glance

Table 1 provides the baseline information on the project.
Figure 5 briefly outlines the process.

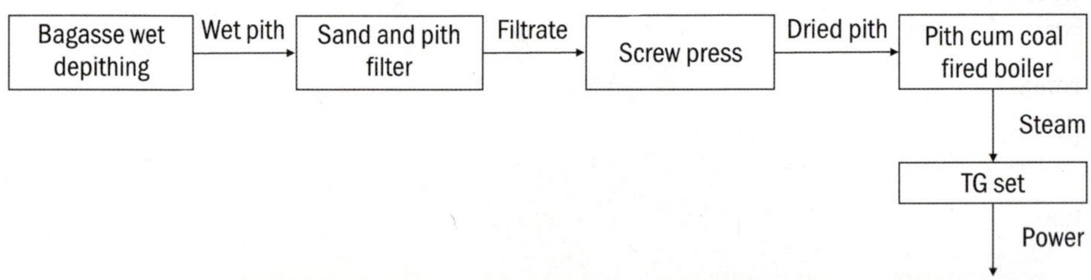

Figure 5 Process description

Environmental impacts

There is no significant impact on the environment. Methane emission that would have occurred in the baseline scenario has been avoided. Consequently, the problem of foul odour in the vicinity of the site where pith used to be disposed of has been addressed. Burning of biomass releases only negligible amount of methane and oxides of nitrogen. There is a marked reduction in the emission of carbon dioxide that was

Table 1 Baseline information

Parameter	Unit	Quantity
Amount of wet pith generated per annum	MT	48 000
Solid mass in the wet pith generated per annum	MT	4800
Water content of the wet pith	MT	43 200
Quantity of water that is removed through screw press	MT	37 333.33
Quantity of dried pith fired into the pith-cum-coal-fired boiler in the year	MT	10 666.67
Quantity of dried pith that was burnt in the coal-fired boiler before the implementation of the project activity	tonnes per hour	8–10
Quantity of dried pith that is combusted in the pith-cum-coal-fired boiler under the project activity in the year	MT	1 066 667
% moisture content of dried pith	%	45
Quantity of bone dry mass of dried pith that is avoided from disposal for anaerobic decay in the year	MT	5866.67
Net calorific value of dried pith	kcal/kg	2124
Quantity of thermal energy available through combustion of dried pith in the pith-cum-coal-fired boiler under the project activity	TJ	52.17
Emission factor of sub-bituminous type of coal that would have been used in the baseline scenario	tCO_2/TJ	96.1
Baseline emissions due to fossil-fuel-based thermal energy generation	tCO_2	5013.62

released due to the burning of coal. Besides, the project leads to the conservation of coal, a precious natural resource.

During dewatering, some waste water is generated in the bagasse wash plant, which is taken to the ETP for treatment and discharge.

Cost–benefit analysis

The screw press has been installed with an approximate cost of Rs 34.80 million. An agro-waste pith/coal-fired boiler has also been installed at a cost of approximately Rs 199.60 million. The net gain is estimated to be Rs 52.1 million per annum, with a payback period of about 3.7 years.

Benefits

The project has contributed to sustainable development as a whole. It has resulted in the following benefits.

- *Environmental benefits* The project activity leads to the reduction in the emission of greenhouse gases like methane and carbon dioxide. Also, there is less emission of SO_2 and NO_x that are generally emitted during the burning of coal. The problem of foul odour emanating because of the disposal of pith has been addressed.
- *Socio-economic benefits* The project has created scope for improving local economy by creating employment and business opportunities. Also, the technology used would have a high replicability.
- *State-of-the-art-technology* The technology used for harnessing the calorific content of the wet pith for steam generation is an innovative and unique technology, converting waste to wealth, with high replicable potential. This project is an example of the implementation of clean technology, leading to sustainable economical and industrial growth.
- *Economics* The net gain because of the implementation of this project is estimated to be Rs 52.1 million per annum.

CASE STUDY AT A GLANCE

- Century Pulp and Paper, a unit of the top Indian corporate conglomerate B K Birla Group of Companies, is one of the leading integrated pulp and paper plants and was established in 1984 at Lalkua, a township near the Himalayas in the state of Uttarakhand.
- The annual turnover of the company was Rs 8457.2 million in 2007/08.
- In Century Pulp and Paper, bagasse, which is generated in sugar mills, is used as a raw material in bagasse-based pulp manufacturing process.
- The pith of the bagasse used to be disposed of on low-lying land.
- The wet pith obtained from the wet depithing process has substantial calorific value.
- In the project implemented by the company, the wet pith is subjected to dewatering using a specially designed screw press customized for the process to reduce the moisture content of the pith from 90% to 50%.
- The pith so obtained is directly co-fired with coal as a fuel in a new specially designed pith-cum-coal-fired boiler to produce steam, which is used to generate power.
- The project activity leads to the reduction in the emission of greenhouse gases like methane and carbon dioxide. The problem of foul odour emanating because of the disposal of pith has been addressed.
- The net gain because of the implementation of this project is estimated to be Rs 52.1 million per annum.

CASE STUDY 13

Sterlite Industries (India) Ltd

Figure 1
Sterlite Industries (India) Ltd

About the company

The Sterlite Industries (India) Ltd, or SIIL, the chief operating company of the Vedanta Resources Group (formerly known as Sterlite Group), is one of the key core sector industries. It was incorporated in 1975 as Rainbow Industries Ltd and was renamed as SIIL in 1986 (Figure 1). It was the first private sector company in India to set up a copper smelter and refinery in the country and has the largest capacity to produce continuous cast copper rods. Currently, it is operating at a rated capacity of 1200 TPD (tonnes per day) of copper. The company is also engaged in captive copper mining on a small scale, though its principal business remains smelting and refining. Apart from its operations in India, the company owns two copper mines in Australia. The SIIL produces finished copper in the form of cathode, some of which is then converted to copper rods. The company is among the fastest growing groups in India, as reflected from its annual turnover, which increased from Rs 500 million in 1997 to Rs 126 720 million in 2007.

Environmental profile

The environment component of the company's combined Health, Safety, and Environment Policy, at both corporate and facility levels, lays emphasis on sustainable development by adopting environment-friendly practices and eliminating adverse environmental impacts. The company is committed to conserving natural resources, raw materials, water, and energy through process improvements and recycling and reducing waste.

The SIIL, Tuticorin, is ISO 9001: 2000, ISO 14001: 2004, OHSAS 18001: 2007, and ISO/IEC 17025: 2005 certified.

Air pollution

The major pollutants emitted into the air include SPM (suspended particulate matter), fugitive gases, and SO_2 (sulphur dioxide).

Air pollution control measures

To reduce air pollution, Sterlite has taken many proactive measures by continually improving process technologies. Some of these measures are listed below.

- Hygiene ventilation system is used for scrubbing fugitive gases.
- Conventional hoods have been replaced with water cooled primary and secondary hoods, which are more efficient in controlling fugitive emissions.
- Secondary scrubbing system is used for scrubbing converter secondary gases, comprising mainly SO_2 (Figure 2).

Figure 2 Secondary gas scrubbing system

- To control fluorine emissions in the phosphoric acid plant, a five-stage scrubbing system has replaced the three-stage scrubbing system.
- To control dust pollution, sides of the conveyor have been covered.
- A new converter bed and heat exchanger have been installed adjacent to the existing converter bed of the H_2SO_4 (sulphuric acid) plant, which has resulted in the reduction of SO_2 emissions.

- The tail gas scrubber system installed in the H_2SO_4 plants has reduced the SO_2 emissions from the stacks.
- Nine ambient air quality monitoring systems have been set up to monitor air quality in the vicinity of the facility (Figure 3).
- Online analysers have been installed in the stacks to monitor SO_2, dust, NO_x (oxides of nitrogen), and CO (carbon monoxide).
- Continuous ambient air quality monitoring systems with online display at gate have been installed at various locations in the facility (Figure 4).
- To control road dust emission, a road cleaning machine is used to clean roads, which uses water sprinkling mechanism.
- Turbo air ventilators have been installed at the refinery to ensure fresh air ventilation.
- An acoustic enclosure has been provided at the steam turbine generator area to reduce the noise level.

Figure 3
Ambient air quality monitoring station

Figure 4
Continuous ambient air quality monitoring station

Water pollution and its control measures

Sterlite is a 'zero discharge' facility since its inception, as the entire treated water is recycled for various plant applications.

Hazardous waste

The different types of hazardous waste generated from the facility include ETP (effluent treatment plant) cake, scrubber cake, and ETP slime. These are disposed of in secured landfill approved by the state pollution control board. Besides, waste oil is also generated, which is sold to the CPCB (Central Pollution Control Board)-authorized recyclers.

Other waste

Other types of waste generated from the facility are granulated slag, lime grit, and gypsum. Granulated slag is used in shot blasting, cement industries, and road construction. Lime grit is recycled back to the smelter, while gypsum is used as a raw material in the cement industry.

Case study

Reduction in hazardous cake generation through process improvement—using sodium sulphide technology

Gas cleaning plant

Sterlite operates a copper smelter, which comprises an Isasmelt furnace, PS (Peirce-Smith) converters, and two H_2SO_4 plants. The SO_2 gas generated in the furnace and converters is treated in the H_2SO_4 plant to produce H_2SO_4 by-product.

The SO_2 gas contains metallic dust, fume, acid mist, water vapour, and various impurities like halides, which may interfere with the catalytic conversion process and contaminate the final product. Therefore, SO_2 is cooled and cleaned in the GCP (gas cleaning plant), which comprises a gas cooling tower, a Dynawave scrubbing system, and wet electrostatic precipitators. A portion of the weak acid from the circulation system of the scrubbers is continuously bled, which is sent to the ETP for treatment. The wet gas precipitator electrically removes the acid mists formed during the downstream operations, and optically fully saturated SO_2-bearing gases are sent to the H_2SO_4 plant.

Ferric sulphate treatment process

About 500–600 m^3 of effluent generated daily from the GCP used to be treated in ETP-2 and ETP-3. These ETPs used ferric sulphate treatment process in which arsenic was removed from the water and fixed in the form of stable ferric arsenate, a hazardous waste. The effluent from the GCP was transferred to the surge (holding) tank. From the surge tank, the effluent was pumped to Reactor-1, where lime slurry was added to maintain a pH of 5.5. At this pH, ferric sulphate was added, which reacted with soluble reactants to produce stable ferric compounds and gypsum. The overflow from the Reactor-1 went to the Clarifier 1. Flocculent was added in the launder to aid flocculation and agglomeration of the fines generated during the chemical reaction in the Reactor-1. The underflow from the Clarifier 1 was diverted to the filter feed surge tank for filtration in pressure filters. The cake from the pressure filter was disposed of in a secured landfill. The filtrate was taken to the Reactor-1 for further treatment.

$2\ H_3AsO_4 + Fe_2(SO_4)^3 \rightarrow Fe_2AsO_4 + 3H_2SO_4$
$H_2SO_4 + Ca(OH)^2 \rightarrow CaSO_4 + H_2O$

The second stage reactor tank content reacted with lime slurry to complete the reaction process at pH 8. The second stage clarifier had the same function as Clarifier 1. The underflow (calcium arsenate) from this was recycled back to the Reactor-1, and the overflow (treated effluent) was pumped to the final holding tank. Figure 5 gives the process detail of ferric sulphate treatment in the ETP-2.

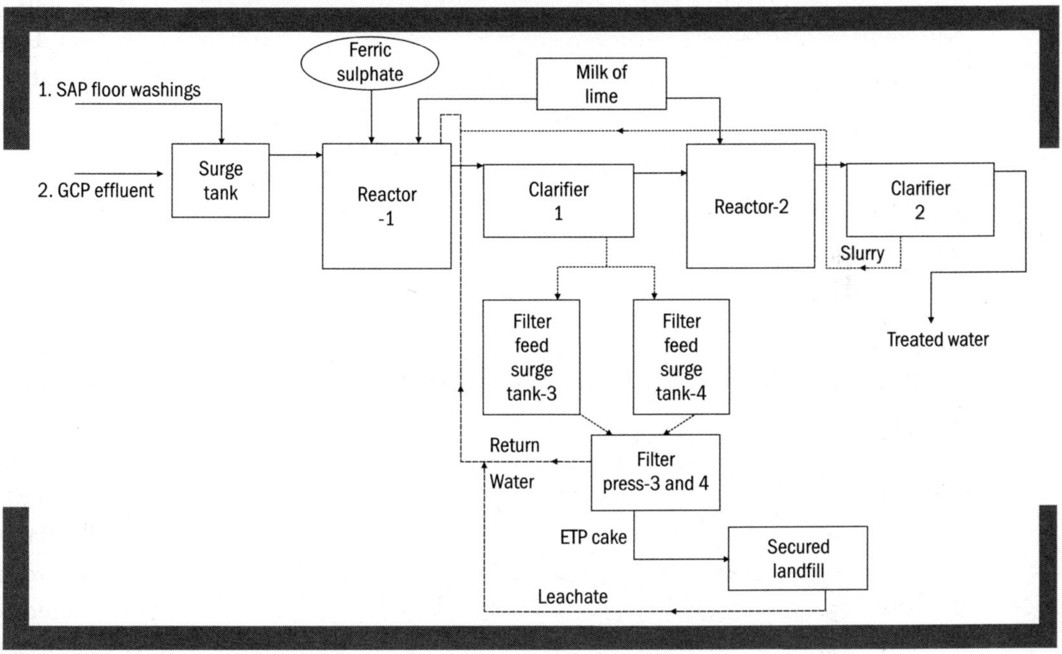

Figure 5 Ferric sulphate treatment process in the effluent treatment plant 2

Sodium sulphide treatment process

To reduce the quantity of hazardous waste generated in the ferric sulphate process (0.25 MT/m^3), the company adopted the sodium sulphide treatment process by installing a new ETP, ETP-4 (Figure 6). The process is based on the affinity of arsenic towards the sulphide ion and low solubility of the product at low pH conditions. This principle is used for the preferential removal of arsenic and other heavy metals from the acidic effluent. According to the principle, when sodium sulphide is added to the acidic effluent containing arsenic and other heavy metals, following reactions take place.

$Na_2S + H_2SO_4 \rightarrow H_2S + Na_2SO_4$
$As_2O_3 + H_2S \rightarrow As_2S_3 + H_2O$
Heavy metals (Cu, Cd, Pb, Se, Fe, and so on) + $Na_2S \rightarrow$ Sulphide of heavy metal + Na_2SO_4

Figure 6 Sodium sulphide treatment in the effluent treatment plant 4

The sulphide precipitation is controlled by pH. This process has reduced the hazardous cake generation from 0.25 MT/m^3 of the effluent treated to 0.11 MT/m^3 of the effluent treated. The GCP bleed from the smelter is treated using sodium sulphide. The heavy metals present in the weak acid effluent are precipitated as metal sulphides. The remaining weak acid free from trace heavy metals is neutralized with lime to produce gypsum, which is non-hazardous in nature. Figure 7 gives the process detail of sodium sulphide treatment in the ETP-4.

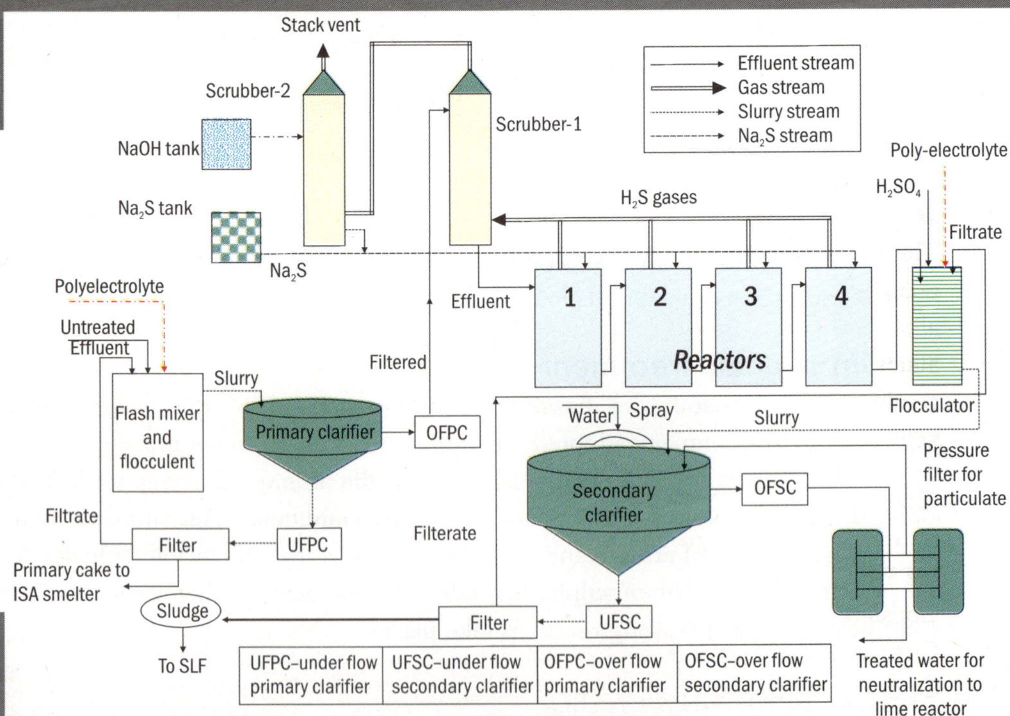

Figure 7 Sodium sulphide treatment process in the effluent treatment plant 4

The major processes/stages involved in the ETP-4 are as follows.

Primary filtration

The effluent from the GCP goes to the surge holding tank. The untreated effluent from the surge holding tank goes to the flash mixer, where polyelectrolyte is added for settling suspended solids in the effluent. The effluent then goes to the primary clarifier, and then to the filtering system through the underflow tank of the primary clarifier. From the filter, the filtrate returns to the flash mixer and flocculator, and the cake generated, which contains approximately 4%–5% copper, is recycled back to the smelter. The overflow from the primary clarifier goes to reactors through the Scrubber-1 (effluent scrubber) to utilize the H_2S gas from the reactors.

Sulphidation process

Sulphide precipitation takes place in three reactors placed in a series. Sodium sulphide is added to all the closed reactors to precipitate the insoluble metal sulphides in the acid. The gases and vapours released from these closed reactors are diverted to scrubbers. After reaction, the slurry goes to the flash mixer, where polyelectrolyte is added. The thickened slurry from the flash mixer goes to the secondary clarifier. The underflow from the secondary clarifier is filtered. The cake generated is disposed of in the secured landfill, and the overflow goes to the pressure filter, where fine particles are filtered.

Neutralization process

The output from the sand filter goes to the lime reactor for neutralization, and the slurry from the reactor goes to the rotary vacuum drum filter for filtration, through the lime clarifier. Clean gypsum is generated from this process. The filtrate from the drum filter is returned to the lime reactor, and the overflow (treated water) from the clarifier goes to the final holding tank of treated water.

Scrubbers

The excess trace amount of H_2S gas is scrubbed in the effluent scrubber by using effluent going to the reactors. The effluent reacts with the H_2S gas, which is then diverted to the alkali scrubber. The resultant effluent is drained to the Reactor-1 for further reactions. In the alkali scrubber, the effluent-scrubbed gases are scrubbed using dilute sodium hydroxide. The resulting scrubbed liquor (Na_2S) goes to the reactors for treatment. The scrubbed gases are vented out through a vent.

Cost–benefit analysis

To treat 1 m^3 of the effluent

ETP-2 or ETP-3

Consumption of ferric sulphate: 0.14 MT = Rs 910

Consumption of lime: 0.03 MT = Rs 180

Operating and maintenance cost, including manpower: Rs 92

Electricity: 2 kWh = Rs 10

Total = Rs 1192

ETP-4

Consumption of sodium sulphide: 0.07 MT = Rs 385
Consumption of lime: 0.015 MT = Rs 90
Operating and maintenance cost, including manpower: Rs 47
Electricity: 20 kWh = Rs 88
Total = Rs 610

Payback on ETP-4

Capital cost: Rs 56.5 million
Operating cost per year: Rs 52.7 million
Cost saving per year to treat 240 m^3/day: Rs 50.3 million
Cake disposal cost saving per year (20 000 MT/year): Rs 8 million
Revenue from non-hazardous gypsum (2400 MT/year): Rs 960 000
Annual saving: Rs 59.26 million
Payback period: 1 year

Benefits

- *Reduction in hazardous waste generation* There is more than 50% reduction in hazardous waste generation after the adoption of sodium sulphide process for treating effluent.
- *Conservation of raw material* Average ferric sulphate required for treating 1 m^3 of effluent is 0.14 MT, while sodium sulphide required for treating the same quantity of effluent is 0.07 MT, that is, 50% less. Similarly, lime consumption has also reduced by half in sodium sulphide treatment process compared to ferric sulphate treatment process.
- *Conservation of land* Generation of less hazardous waste means less waste going to landfills, thus leading to land conservation.
- *Saving* By using sodium sulphide treatment process, annual saving to the tune of Rs 59.26 million is achieved.

CASE STUDY AT A GLANCE

- The SIIL, the chief operating company of the Vedanta Resources Group (formerly known as Sterlite Group), is one of the key core sector industries dealing with copper mining, smelting, and refining.
- About 500–600 m^3 of effluent is generated daily from the gas cleaning plant, which used to be treated in the ETP using ferric sulphate treatment process, in which arsenic was removed from the water and fixed in the form of stable ferric arsenate, a hazardous waste.
- To reduce the quantity of hazardous waste generated in the ferric sulphate process, the company adopted the sodium sulphide treatment process.
- There is more than 50% reduction in hazardous waste generation after the adoption of sodium sulphide process for treating effluent.
- Average ferric sulphate required for treating 1 m^3 of effluent is 0.14 MT, while sodium sulphide required for treating the same quantity of effluent is 0.07 MT, that is, 50% less. Similarly, lime consumption has also reduced by half in the sodium sulphide treatment process compared to the ferric sulphate treatment process.
- Generation of less hazardous waste means less waste going to landfills, thus leading to land conservation.
- By using sodium sulphide treatment process, annual saving to the tune of Rs 59.26 million is achieved.

SECTION

III

Process improvement and product innovation

The industrial sector uses about 50% of the total commercial energy available in India. Coal, lignite, and oil and natural gas are the main sources of energy being used. Of the commercial sources of energy, coal and lignite account for about 56% of the total energy used; oil and natural gas, around 40%; hydroelectric power, about 3%; and nuclear power accounts for 1% of the total amount of energy used. A majority of the industrial sector is highly energy intensive, and the efficiency is well below in comparison to other industrialized countries. The Energy Conservation Act, 2001, has identified designated consumers that have a large potential for saving energy, such as cement, aluminium, paper and pulp, iron and steel, textiles, chloro-alkali, thermal power station, railways, and fertilizers. It has been estimated that the energy-saving potential of these industries is approximately 25%.[1] About 5% to 10% savings can be realized by adopting best operating practices. Low/medium investment on various energy conservation measures, such retrofitting and use of energy-efficient devices and controls, would help in 10%–15% savings. Major breakthrough in savings is possible with process modifications, process improvement, as well as product innovation. This section presents case studies demonstrating successful initiatives undertaken by different organizations to optimize energy utilization and reduce adverse impact of industrial activities on the environment.

---

[1] Hand book on energy audit and environment management. p. 2, TERI 2006

CASE STUDY 14 Ayurvet Ltd

Figure 1
Ayurvet Ltd, Sonepat

About the company

Ayurvet Ltd, formerly known as Dabur Ayurvet Ltd, is one of India's leading animal care companies, specializing in 100% natural and safe herbal products. It is the veterinary product division of Dabur India Ltd and was founded in 1992. Its products are based on the philosophy of ancient knowledge of *Ayurveda*, the oldest surviving complete medical system in the world. The company blends ancient knowledge of *Ayurveda* with modern precision and technology to deliver scientifically researched quality formulations for animal health care. The company's Animal Feed Plant is located in Chidana in Haryana and manufacturing facility is situated in Baddi (Figures 1 and 2). It brings out a range of products for cattle, poultry, pets, pigs, aquamarine animals, and equine. The turnover of the company in 2007/08 was Rs 325 million.

Figure 2
Ayurvet Ltd, Baddi

Environmental profile

The company has an Environment Policy at both corporate and facility levels and an environment department. The AGES (Ayurvet Green Energy Solution) addresses the environmental issues, and innovates and promotes environment-friendly technologies. The company does not emit any air pollutant or generate any water pollutant or hazardous waste.

Case study

Development of technology for hydroponics fodder producing machine and system

Hydroponics

The word 'hydroponics' has been derived from the Greek word meaning 'water working', which essentially means growing plants without soil. Hydroponics was practised many centuries ago in Amazon, Babylon, Egypt, China, and India. The 'Hanging Garden of Babylon' and the Aztec's floating farms were actually prototypes of hydroponic systems. Later, plant physiologists started growing plants with specific nutrients for experimental purposes, naming this practice as nutriculture. In 1929, Dr William F Gericke of the University of California succeeded in growing tomato vines of 7.5 metres height in nutrient solution. He named this new production system hydroponics. The practice of growing crops in inert solid media using nutrient solution is also termed hydroponics.

In addition to developing hydroponic systems for the production of vegetables, work started between 1930 and 1960 to develop a system to produce poultry feed. The system comprises greenhouses and protective covers and is called controlled environment agriculture. Researchers found that cereal grains could be grown very rapidly in this manner.

To make available green fodder year-round, there is a need to develop technologies that are farmer-friendly, sustainable, and economically viable. Ayurvet Ltd and R K Goel Institute of Technology joined hands to develop a hydroponic fodder production machine and system. The programme formally started in January 2007 and got over in December 2007. The hydroponic machine, with a production capacity of 50 kg/day, has been installed at the premises of the R K Goel Institute of Technology, Ghaziabad. Two models with 500 kg biomass production/day and 1000 kg biomass production/day would be installed at the Ayurvet Feed Plant at village Chidana in Sonepat district, Haryana.

Conditions

Controlled environment agriculture using hydroponics for growing fodder in dry and arid regions is attractive, provided growth of fungi is contained even at a temperature of 30–32 °C and at a relative humidity of 80%–85%. Also, ingress of heat should be avoided without affecting the photosynthesis-friendly solar spectrum. Further, optimum range of humidity and temperature will be found in the experimental fodder chamber under the following conditions.

- Power and water requirement is minimal.
- Cleanliness of environment is ensured.
- Chemicals like chlorine are not used.

Process/apparatus

The process requires a closed insulated chamber, where ozone is used to keep the air and water clean. Products are grown in multilayer shelves. Controlled light is let in through suitably glazed windows. Water is recirculated and sprinkled at intervals. Soaked seeds and sprouts are kept in shelves daily or at regular intervals, depending on the product, and harvesting can be done daily or at suitable intervals. In this project, after studying the variation in temperatures in summer, winter, and monsoon months in and around Delhi, it was concluded that evaporative cooling will be required during the hot and dry months of May and June. Heating will be required for few hours at nights during January–March and October–December. During April, July, and August, air circulation is sufficient to maintain desirable environment.

To get maximum benefit from hydroponic fodder, it needs to be used in conjunction with dry matter. Under controlled conditions, each kg of grain will produce about 8 kg of hydroponic fodder within seven to eight days. Hydroponic feeds are 95%–98% digestible, unlike sprouted grains, which are at best 30% digestible.

Based on the laboratory experiments, a hydroponic chamber was designed and fabricated. It has successfully worked for the last several months in the temperature range of 15–32 °C. Ozonated air and water were used to inhibit fungi.

The apparatus used consists of foldable chambers and is made up of variety of materials, including wooden boards, aluminium/GI/iron/plastic sheets with or without insulation, brickwork, and concrete. Multilayered shelves are created in the chambers.

The hydroponic machine has been installed at the premises of the R K Goel Institute of Technology, Ghaziabad, for demonstration and further refinement of the technology. The production capacity of this prototype is 50 kg biomass production per day. The experimental chamber will be scalable to higher production rates of the order of 1000 kg biomass per day (Figures 3 and 4).

Figure 3
Ayurvet hydroponics machine

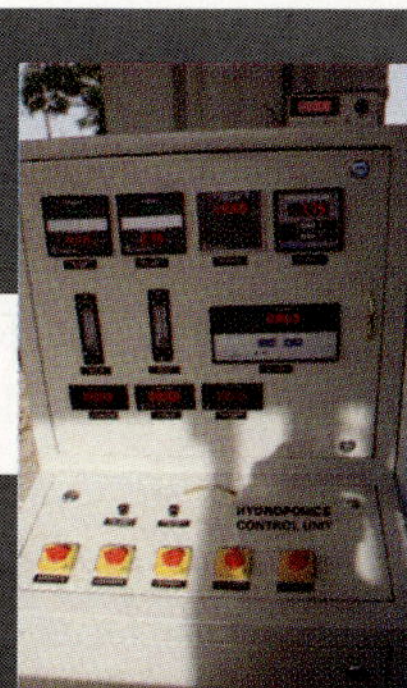

Figure 4
Hydroponics control panel

Results

- Under the conditions of 34 °C temperature and 85% relative humidity, fodder was successfully grown.
- Growth of fungi was controlled by using ozonated water for washing seeds, standardization of ozonated water spray in the growth chamber, and standardization of ozonation of air in the chamber.
- In some experiments, for short duration of three to four hours, temperatures were allowed to rise up to 38 °C successfully.
- The grains used in the study were maize, wheat, barley, paddy, corn, and so on.
- Fodder conversion rate was of the order of eight to nine times (Figure 5).
- Ayurvet nutrient solution-1 was used as the nutrient, as it gave the best results.

Figure 5
Hydroponics fodder

Cost–benefit analysis

The project was operational for one year and it incurred a cost of Rs 650 000.

Benefits

- Hydroponics technology was developed using low-cost technology, in which no soil was used.
- While globally, farmers grow fodder using this technology at a temperature of 20–22 °C, this indigenous technology makes cultivation possible at a higher temperature.
- The hydroponic chambers used worldwide use artificial lights, but this indigenous technology uses natural light only.
- This technology saves land, as production of 1000 kg hydroponic fodder requires 300 square feet of land only.
- Considerable amount of water is saved (water usage is 2–3 litres/kg in the hydroponic system as against 80 litres/kg in the conventional system).
- This technology saves time (it takes six to seven days for hydroponic fodder production as against 45 days taken (for growing maize) in the field).

- The yield in the hydroponic technology is more compared to the conventional method.
- The hydroponic system works indoors and is less arduous.
- For a 1000 kg/day production, 4 man hours/day are required.
- Contamination-free hydroponic fodder is grown in aseptic conditions, while conventional fodder is affected by several pests and diseases.
- There is a constant feed supply, as hydroponic fodder can be harvested 359 days in a year, initial time gap being seven to eight days.
- The fodder produced in the hydroponic system is highly digestible and more nutritious.
- fodder produced using hydroponic technology improves animal health.

CASE STUDY AT A GLANCE

- Ayurvet Ltd, formerly known as Dabur Ayurvet Ltd, is one of India's leading animal care companies, specializing in 100% natural and safe herbal products.
- The turnover of the company in 2007/08 was Rs 325 million.
- Ayurvet Ltd and R K Goel Institute of Technology joined hands to develop a hydroponics fodder production machine and system. The programme formally started in January 2007 and got over in December 2007.
- The project incurred a cost of Rs 650 000.
- The production capacity of this prototype is 50 kg biomass production per day.
- The grains used in the study were maize, wheat, barley, paddy, corn, and so on.
- Fodder conversion rate was of the order of eight to nine times.

CASE STUDY 15

Tata Steel

Tata Steel

About the company

Established in 1907, Tata Steel is Asia's first and India's largest integrated private sector steel company (Figure 1). It is the world's sixth largest steel company, with an annual crude steel production capacity of 28 MT (million tonnes). Ranked 315th on Fortune Global 500, it is based in Jamshedpur, Jharkhand, India. The company has set an ambitious target to achieve a capacity of 100 MT by 2015. Tata Steel has also lined up a series of Greenfield projects in India and outside.

Crude steel is the major product manufactured at Steel Works, Jamshedpur. Its production in 2007/08 was 5 014 000 tonnes. The turnover of the company in 2007/08 was Rs 1321.1 billion.

Environmental profile

The Environment Management System, or EMS, at the Steel Works, Jamshedpur, was established in 1998, and the ISO 14001:1994 Certificate was granted in 2000. Subsequently, the system was upgraded to ISO 14001:2004 version and recertified in 2006. The EMS was integrated with Occupational Health and safety Management System, or OHSMS, in 2002. Under the EMS, systems were implemented with the following objectives.

- Monitoring and tracking of pollution control equipment
- Water surveillance to prevent water losses
- Adoption of best management practices

Initiatives

Tata Steel's other initiatives in the field of environment are listed below.

- Adopting the Tata Code of Conduct for business, which addresses the environment, health, safety, ethics, and other elements of business practices.
- Formulating policies for the environment.
- Closing down the old and polluting units. Till date, 2 MT capacity units have been shut down.
- Committing to reduction in CO_2 (carbon dioxide) emissions.
- Adopting environment-friendly state-of-the-art-technologies during expansion and modernization.
- Implementation of Total Quality Management Practices for operational efficiency.

Environmental risk management and disaster management plans

Tata Steel has prepared and approved the Onsite Emergency Plan and Disaster Control for Steel Works. This plan was prepared in accordance with the Factory Act, 1948, Control of Industrial major Accident hazardous Rules, 1990, and MSI of Hazardous Chemicals Rules.

Air pollution

Air pollution is characterized by the emissions of SPM (suspended particulate matter), RSPM (respirable suspended particulate matter), SO_2 (sulphur dioxide), NO_x (oxides of nitrogen), along with fugitive emissions.

Air pollution control measures

SPM, RSPM, and fugitive emissions

The various combustion processes taking place in the Steel Works are mainly responsible for the emissions of PM (particulate matter). Stock piles, raw material handling, vehicular emission, and quenching processes also add to the PM load. The facility has implemented state-of-the-art technologies and latest pollution control equipment to control stack emissions. These are listed below.

- ESPs (electrostatic precipitators) to control SPM emission
- Bag houses
- Dust suppression systems in the form of water spray, dry fog, and chemicals at conveyors and material handling units.
- Secondary fume extraction systems at LD (Linz–Donawitz) process shops.
- Scrubbers and wet cleaning plants at blast furnaces and LD shops.

Apart from these, roads are swept and cleaned regularly, and more than 50 000 trees have been planted in the premises to control dust.

On account of these control measures, the PM load from stacks has shown marked reduction over the years. It has reduced by 43% over the past five years (Figure 2).

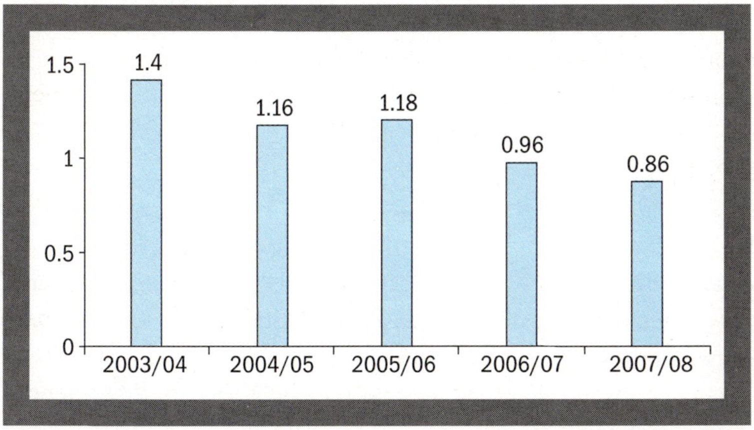

Figure 2 Stack emission from 2003/04 to 2007/08 (PM-kg/tcs)

SO_2 and NO_x emissions

The combustion processes taking place in coke ovens, sinter-making units, blast furnaces, finishing mills, and power houses release SO_2 and NO_x. Following measures are taken to reduce SO_2 and NO_x emissions.

- Following efficient combustion process.
- Reducing the consumption of fossil fuels.
- Using low– NO_x burners.

With the adoption of these strategies and control, the Steel Works has managed to bring down its emission load significantly over the years, as shown in Table 1.

Table 1 Emissions (tonnes per annum) at the Steel Works, Jamshedpur

Pollutant	2005/06	2006/07	2007/08
Particulate matter	5570	4666	4318
Sulphur dioxide	7287	6639	4669
Oxides of nitrogen	6215	8264	5994

CO_2 emission

The Steel Works is aware of the fact that CO_2 is a potent GHG (greenhouse gas), contributing to global warming in a big way. So the facility has taken various initiatives to reduce its emission, as listed below.

- Waste heat recovery from blast furnace stoves.
- Conversion of all coal-fired boilers to by-product gas fired boilers.
- Retrofitting of coke dry quenching facility at existing coke batteries (under process).
- Top recovery turbine for G and H blast furnaces (under construction).
- Installation of variable frequency drives at various points in the plant.

Noise pollution control

Noise levels in the facility are controlled by adopting following measures.

- Using machinery that generates less noise.

- Moving to technology that does not cause noise pollution.
- Using silencers/mufflers in the chimneys of diesel generators.
- Vibration dampening anchoring.
- Designing a room keeping acoustics in mind.
- Controlling steam leakages.

Water pollution

The main water pollutants discharged by the facility are BOD (biochemical oxygen demand), COD (chemical oxygen demand), pH, oil and grease, TSS (total suspended solids), TDS (total dissolved solids), heavy metals, CN (cyanide), and ammonia.

Water pollution control measures

A combination of best available physico-chemical and biological processes is used for treating the waste water generated from the facility. The following list gives an overview of various treatment methods adopted at the facility.

- Waste water from the coke plant is treated biologically, wherein organic pollutants are subjected to oxidation and decomposition by micro-organisms.
- Treatment plants have been installed for casters
- Effluents from the power plant are neutralized.
- Cooling tower blow down water treatment plants have been installed.
- Cole rolling mill effluent is treated in biological treatment plants.
- Facilities for removing oil and mill scale have been provided in various mills.

Hazardous waste

The Steel Works generates hazardous waste in the form of tar sludge, BOD sludge, CN sludge, zinc dross, chrome sludge, lube oil/coolant oil sludge, lead acid batteries, and used lube oil/grease. These waste constituents are disposed of in the following manner.

- Tar sludge and BOD sludge are recycled to the coke plant.
- CN sludge is treated in the BOD plant.
- Zinc dross is sold to authorized agencies.
- Chrome sludge is landfilled in protected landfills.
- Lube oil/coolant oil sludge is recycled to the coke plant or power plant.
- Lead acid batteries are auctioned to registered recyclers.
- Used lube oil/grease is recycled to the blast furnace.

Other waste

Apart from the above-mentioned pollutants and waste, the facility generates many types of process waste like blast furnace slag, LD slag, blast furnace sludge, LD sludge, mill scale, flue dust, lime fines, dolo dust, mill sludge, and so on. Most of these are utilized in various processes. For example, LD slag, LD sludge, mill scale, lime fines, dolo dust, U/S limestone, and mill sludge are used in the sinter-making process. Blast furnace sludge and flue dust are used as domestic fuels. The facility utilized an impressive 85.43% of waste in 2007/08. Utilization of LD slag still remains a challenge, and it forms a major portion of unutilized waste.

Case study

Elimination of coal firing at power house-3

Objectives

- Eliminating the burning of coal for power generation.
- Recovery of waste heat from by-product gases, and using it for power generation.
- Conversion of old stoker-fired boilers into gas-fired boilers.

Background

Power House-3, or PH-3, is one of the oldest units at the Steel Works, constructed before 1950 (Figure 3). The PH-3 used to meet substantial captive power requirement, to the tune of 30–45 MW, of the facility. It comprised the following units.

Figure 3
Power house

- Four coal-fired spreader stoker boilers (boiler numbers 5, 6, 7, and 8) with steam generating capacity of 67.5 TPH (tonnes per hour) each at 485 °C temperature and 63 bar (g) pressure.
- Two condensing steam turbine generators of 30-MW capacity each.
- Coal handling, ash handling, and other auxiliary facilities.

Environmental impacts of using a fossil fuel

Burning of middling coal, a fossil fuel, in the power plant used to emit 450 000 tonnes of CO_2 per annum, apart from other pollutants like dust (48 kg/h), SO_x (oxides of sulphur; 242 kg/h), and fly ash and bottom ash (213 000 tonnes per annum).

Generation of waste gases

The integrated steel plant of Tata Steel generates BFG (blast furnace gas), COG (coke oven gas), and BOFG (basic oxygen furnace gas) from its operating units. A part of these gases is used in the steel plant. However, the surplus gases, after cleansing, were being flared into the atmosphere, as a result of which the heat content of these gases was getting dissipated. The process followed at PH-3, prior to gas firing, is depicted in Figure 4.

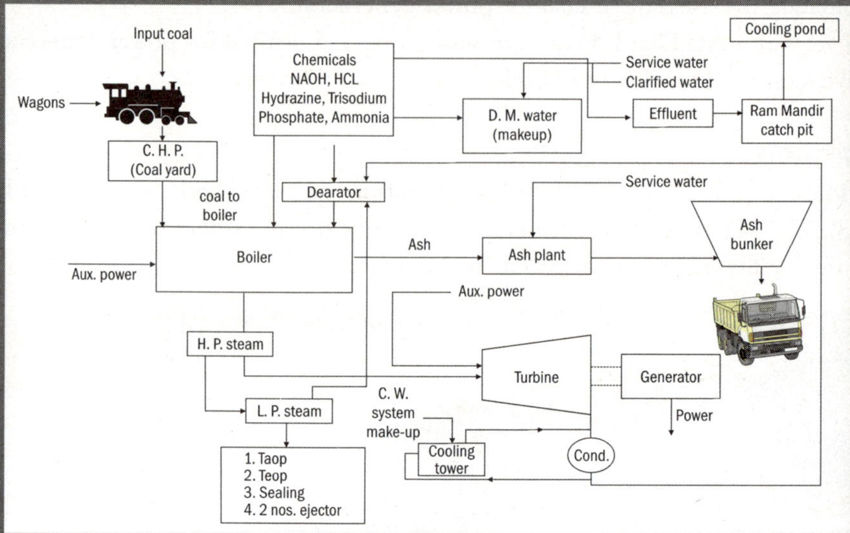

Figure 4 Process at PH-3

Proposal to utilize the surplus waste gases

To prevent the wastage of useful heat of these gases, it was proposed that this heat be utilized for generating electricity. Thus, the project activity entailed the recovery of thermal energy potential of the surplus waste gases and using it at the PH-3 for generating steam, and subsequently, electricity. To achieve this goal, it was necessary to convert all existing coal-fired boilers of the PH-3 to 100% waste gas fired boilers. The electricity generated at the PH-3 met the critical load power requirement of the Steel Works. Therefore, it was necessary to maintain the same level of generation after the implementation of the project.

Modifications in the boilers

Conversion of coal-fired boilers of the PH-3 to 100% waste gas fired boilers required some major modifications in each boiler. This involved installation of technologically advanced tri-fuel scroll burners in the boilers, capable of firing leans gases like BFG and BOFG, richer gas like COG as a support fuel, and LDO (light diesel oil) as a secondary fuel. Besides, some structural modifications were also made. The heat transfer from the waste gases was mainly by means of convection, which requires a large surface area in heat exchanger, for example boiler. So, the surface area of the boilers was enhanced. The panels related to stroker drives, feeders, coal handling, ash handling systems, and ESPs, and unwanted infrastructure were dismantled. A distributed control system with engineering station, human–machine interfaces, servers, upgraded instrumentation, control platform, and a suitable combustion control system were introduced. Conversion of boilers started in 2006, and all the boilers were successfully converted by 2007 (Figures 5 and 6).

Figure 5
Boiler before conversion

Figure 6
Boiler after conversion

Transportation of gases

A gas pipeline network (<2 km length) was established to transport waste gases from their source of generation to the PH-3 (Figure 7). This was quite a challenging task, considering the distance and the volume of the gas to be transported.

Challenges faced

Challenge

Achieving the proper heat balance in converting the stoker-type coal-fired boilers into gas-fired boilers.

Figure 7
Gas transportation from the source to the PH-3

Actions taken

- Increasing the heating surface area by increasing the number of tubes.
- Modifying front walls and super heaters.
- Replacing boiler tank tube in the flue.
- Providing additional air pre-heaters.

Challenge

Management of steam and meeting power requirements of the Steel Works during the conversion period.

Actions taken

- Carrying out boiler conversion in stages.
- Applying critical chain project management tools for implementing the project.

Challenge

Supplying waste gases to the PH-3, located more than 2 km away from the gas generating source, at required pressure.

Actions taken

- Developing control philosophy for supplying gas at higher pressure so that it is available at the right pressure at the boilers of the PH-3.
- Laying more than 2-km long overhead DN2400 gas pipeline for transporting gases.

Benefits

After the conversion of the boilers at PH-3 to waste gas fired boilers, there has been an improvement in efficiency (Figure 8), and the concentration of pollutants generated has reduced considerably, as depicted in Table 2.

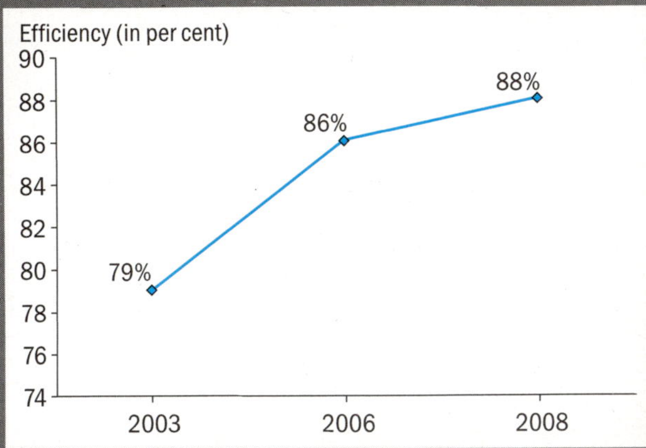

Figure 8 Trend in the boiler efficiency over the years

Table 2 Emissions of various pollutants before and after conversion

Consumption/emission	*Unit*	*Value before conversion*	*Target after conversion*	*Actual achievement*
Coal	tonnes/annum	456 700	Nil	0
Carbon dioxide	tonnes/annum	450 000	60 000	54 236
Bottom ash	tonnes/annum	131 500	Nil	0
Fly ash	tonnes/annum	81 800	Nil	0
Dust	kg/h	48	20	17
Oxides of sulphur	kg/h	242	100	97

CASE STUDY AT A GLANCE

- The 5-MT-capacity Jamshedpur plant produces crude steel, and its turnover in 2007/08 was Rs 1321.1 billion.
- The main aim of the project at PH-3 (power house-3) was to maximize the usage of by-product gases and reduce coal-based steam generation.
- This aim was achieved by the conversion of all existing coal-fired boilers of PH-3 to 100% waste gas fired boilers.
- The power thus generated at PH-3 caters to the critical load power requirement of the Steel Works of Tata Steel.
- Burning fossil fuels (middling coal) used to release 450 000 tonnes per annum of CO_2, apart from other pollutants such as dust, other gases, and ash.
- After the conversion of the boilers, the emissions of all the pollutants either reduced or became nil.
- There has also been an improvement in the overall efficiency of the boilers.

CASE STUDY 16 Shree Cement Ltd

The main plant of Shree Cement Ltd

About the company

Shree Cement Ltd, with its present capacity of 6.8 MTPA (million tonnes per annum), is the largest plant at a single location in North India (Figure 1). The company commenced its operation in 1985 at Bangur Nagar, near Ajmer, Rajasthan, by establishing a 0.6 MT (million tonne) capacity unit. It has expanded its operations remarkably, and it currently boasts of two cement plants at Bangur Nagar, Beawar; four cement plants at Bangur City, Rajasthan, district Pali; and two grinding units at village Khushkhera, near Gurgaon. The company manufactures ordinary Portland cement and red oxide brand cement, with their annual production in 2007/08 being 1 184 772 MT and 1 911 845 MT, respectively. The company also has a captive power generation unit of 101.5 MW capacity. The company's turnover in 2007/08 was Rs 20.66 billion, and it is all set to touch Rs 30 billion mark in 2008/09.

Environmental profile

Environment cell

Shree Cement is an environment conscious company. It has a state-of-the-art environment cell, which comprises a monitoring cell and a maintenance cell. The monitoring cell works in close association with the Rajasthan Pollution Control Board, CPCB (Central Pollution Control Board), Ministry of Environment and Forests, and other government agencies. As the name suggests, the maintenance cell is responsible for the smooth functioning and upkeep of pollution control equipment and machineries. There is also a separate cell for the development and maintenance of green belt in the plant premises.

Environment policy

The company also has an Environment Policy in place. The policy ensures efficient use of natural resources, energy, and equipment; reduction in emissions of GHGs (greenhouse gases), noise, and waste; continual improvement in environment management; and compliance to relevant environmental legislations. In short, the policy aims towards a clean, green, and healthy environment.

Shree Cement is an ISO 14001 (EMS), 9001–2000 (QMS), OHSAS-18001 (OHSMS), and SA 8000 certified company, which further underscores its commitment to the environment and sustainable development.

Air pollution

The major air pollutants emitted from cement plants and mines are SPM (suspended particulate matter) and RSPM (respirable suspended particulate matter). However, a small quantity of NO_x (oxides of nitrogen) is also released during the burning process. Fugitive and CO_2 (carbon dioxide) emissions also form a major chunk of emission.

Air pollution control measures

Stack management

ESPs (electrostatic precipitators) and bag houses have been installed in kiln stacks, cooler stacks, and cement mill stacks. Coal mill exhausts are connected to a high-efficiency bag house. All emitting frames have been replaced with new fabricated frames to increase the efficiency of the raw mill I ESP (Figure 2).

Ambient air quality management

- A separate air line for coal mill I bag house has been set up to address the problem of high differential pressure and increase the efficiency of the dust collector.
- Additional bag filters have been installed. Dust collectors have been provided at the clinker loading area, deep conveyor belt I, and coal transport area.

Figure 2
Electrostatic precipitators

- Old nozzles have been replaced with modified nozzles that spray water in a better and efficient manner at the mine crusher and unloading hopper.
- Indigenous bags have been replaced with imported bags.
- Rock breakers are used for secondary blasting. This has resulted in the suppression of fly rocks, vibrations, and dust generation. Besides, additives and air decking are used in blasting to reduce charge/delay, thereby noxious fumes.
- Vaporized water injection system has been installed in drilling machines to reduce the dust emission.
- Fly ash, mill scale, agro waste, and so on are used in manufacturing blended cement.

Fugitive emission management

- Fugitive dust generated at various transfer points is suppressed by spraying fine droplets of water and through various dust collection bag filters installed at strategic points.
- About 35% of the plant area has been planted with 157 000 plants in a bid to reduce fugitive emissions. Similarly, about 45.1 hectares of the mining area has been brought under plantation.
- Unloading hopper has been provided with an atomized water spray system at mine and limestone crushers and coal unloading points.
- Atomized water spray system has also been installed at limestone stacker, coal stacker, and reclaimer.
- Covered storage has been provided for clinker and gypsum.
- The transport belts used for transporting raw material and finished products are covered.
- The facilities are provided with concrete roads.
- Sweeping machines are used for sweeping and keeping the facilities dust free.

Noise pollution control measures

- In the limestone mining areas, non-electric detonators and top line detonators are used to reduce noise and vibration.
- Drilled detonators are used to suppress noise wave frequency associated with blasting.
- Plant equipment and machinery are properly lubricated and maintained so that they do not generate noise.
- Green belt and high plant boundary wall serve as effective sound barriers.

Water pollution and its control measures

Shree Cement uses dry process technology for cement production. Hence, there is no water pollution during the cement manufacturing process. Water is used for cooling the plant equipment, which is then recycled. Water is also sprayed in the raw material yard, crusher hopper, and on roads to control fugitive emissions. Domestic sewage generated from the company's colony is treated in the uplift flow filter, and this treated water is used to water plants grown in the green belt.

Waste water generated from the captive thermal power plant and waste heat recovery system (Figure 3) is neutralized and is then sprayed on mine roads, coal yard, and raw material yard.

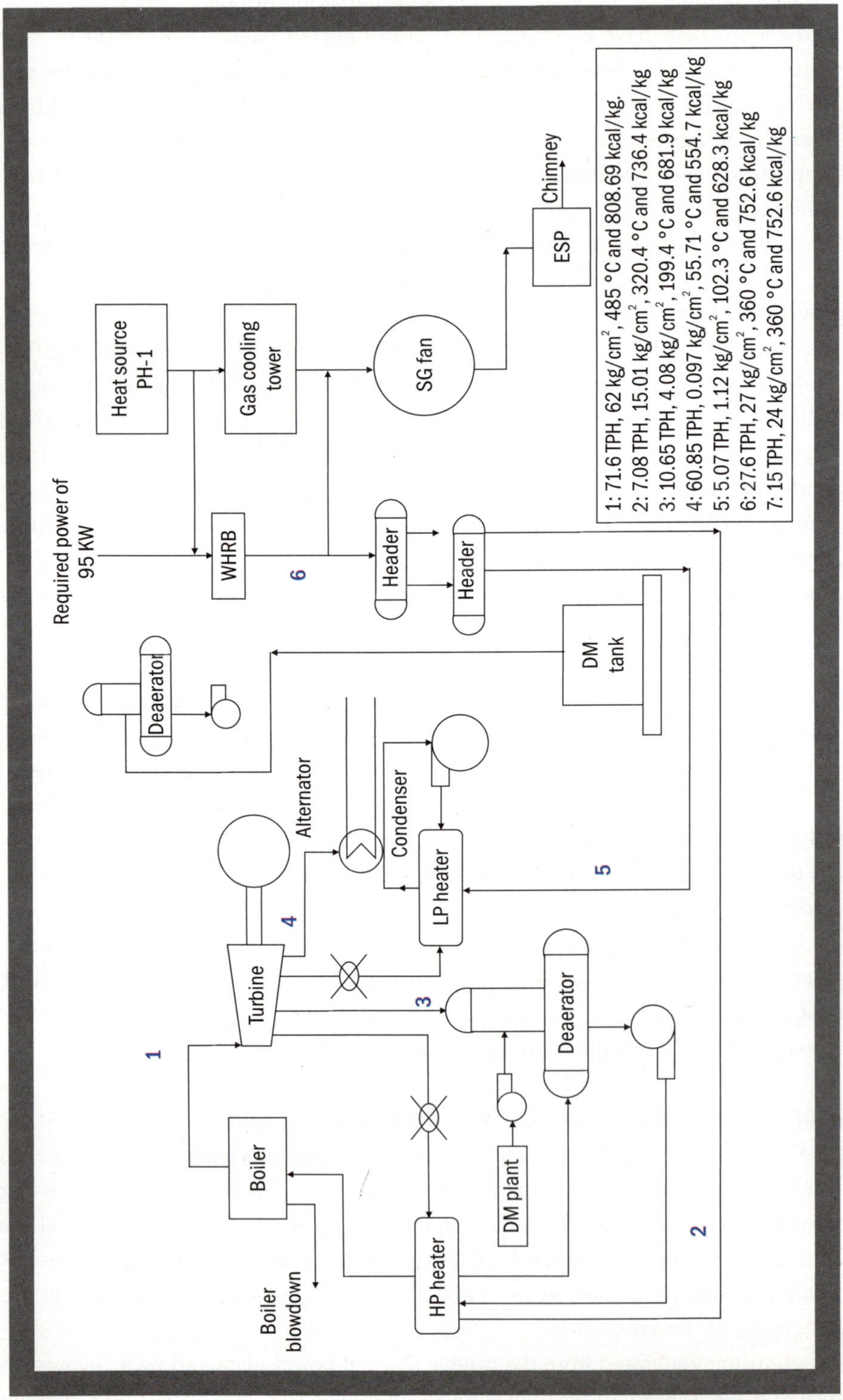

Figure 3 The present waste heat recovery system

Hazardous waste

Used lube oil is the only hazardous waste generated from cement plants and the thermal power plant. Some of this oil is recycled and reused in stacker and reclaimer chain assembly as a lubricant and the rest is sold to the CPCB-authorized dealers.

Other waste

Shree Power Plant generates 100–110 tonnes/day of fly ash and bed ash. The fly ash generated is utilized in making blended cement. Boiler ash generated by Shree Power is utilized as a low calorific value fuel in the process of clinkerization. Thus, the facilities are zero disposal facilities.

Case study

Conservation of natural resources by using flue gases of the cement plant for steam generation in the power plant

Introduction

Shree Cement firmly believes in the adage 'clean and green is profitable'. To make its operations green and sustainable, the company has taken various measures for the conservation of natural resources like water and fuel. These measures are a combination of proper management of resources with energy efficiency practices.

Water conservation

Beawar in Rajasthan, where Shree Cement is located, is a water-scarce region. Although the company uses dry process to manufacture cement, which does not require much water, it has contributed to water conservation in the region by taking many novel initiatives.

Installation of second reverse osmosis system at the power plant

In the power plant, initially, there was one RO (reverse osmosis) plant (Figure 4), RO-I, for the production of boiler feed water. This RO plant used to generate 45 kl (kilolitres)/h reject as waste with high TDS (total dissolved solid) content. With the installation of second RO unit, RO-II, the quantity of waste water generated has reduced to a large extent. Of the 45 kl/h reject water from RO-I, 27 kl/h is further treated in RO-II. The capacity of RO-II is 27 kl/h and the recovery is 67%. Total waste water coming out of this plant is 196 kl/day.

This waste water has replaced fresh water for reducing the temperature of pre-heater exit gases in the gas conditioning tower after incorporating several corrective and remedial measures to address the associated problems. This has resulted in the saving of 196 kl of fresh water per day.

Installation of air-cooled fluid cooler at the power plant

The cooling tower section in the power plant used to witness heavy water loss, to the tune of 3000 kl/day. This huge quantity of water used to get evaporated, leaving behind water with high TDS, which was difficult to dispose of. To reduce this huge loss, an air-cooled fluid cooler was installed. This system is working on 2 kW of electricity, saving 109 kl/day of water. Figure 5 shows the air-cooled fluid cooler.

Figure 4
The reverse osmosis plant

Figure 5
The air-cooled fluid cooler at the power plant

Recycling of domestic waste water

An STP (sewage treatment plant) has been installed for treating the domestic waste water of the residential colony (Figure 6). About 350 kl of treated water is generated every day, which is used for watering plants in the green belt developed in the premises and also for spraying on roads and hoppers, and so on.

Other measures

Shree Cement has installed a roof water harvesting system at the plant level and has constructed check dams in the nearby areas. These measures have contributed significantly towards recharging of groundwater in the area.

Figure 6
The sewage treatment plant

Natural resource conservation in the cement plant

Generation of waste heat

About 90% of the total energy used in the clinker burning process comprises heat energy, which is used for burning limestone at a temperature greater than 1450 °C. Of the total heat used, about 55% is used in the burning process. The rest 45% was discharged as sensible heat through the exhaust gases of the pre-heater. Of this 45% heat, 10% was used for drying the raw material and coal, and the rest 35% was emitted to the atmosphere in combination with pre-heater associated gases. Before these gases were emitted to the atmosphere, they were passed through a GCT (gas cooling tower) and an ESP.

In this process, huge quantity of water was used in the GCT, and heat energy associated with waste gases had no utilization.

Utilization of the waste heat

After conducting a thorough research, the waste heat associated with pre-heater gases in the cement plant is now used for generating steam in the power plant in a waste heat recovery boiler (Figure 7), which is then used for pre-heating the feed water in high-pressure and low-pressure heaters in the plant. The earlier practice of extracting steam from the turbine is now discontinued, leading to saving of fuel. This has resulted in the saving of water, as now there is no need for passing the flue gases through the GCT. The flue gases are passed through the ESP and released into the atmosphere.

Constraints faced

Flue gases contain high concentration of dust, which is sticky, abrasive, and corrosive in nature. This dust used to accumulate on the surface of boiler tubes. Dust collection and dust dislodging systems were installed in succession, but they did not work. Finally, a dust hammering system

Figure 7
Waste heat recovery boiler

has been installed, which is functioning successfully. There are other technical problems related to the waste heat recovery boiler, which are being looked into.

Cost–benefit analysis

Due to installing a waste heat recovery boiler
Savings due to fuel conservation = Rs 112.3 million per annum
Savings due to water conservation = Rs 5.6 million per annum
Total savings = Rs 117.9 million per annum
Savings at 85% load factor = Rs 100.2 million per annum
Annual expenditure on operation and maintenance = Rs 5 million
Net gain on installing the waste heat recovery boiler = 100.2 – 5 = Rs 95.2 million per annum
Total project cost of the waste heat recovery boiler = Rs 192.2 million
Payback period = 192.2/95.2 = 2.01 years

Benefits

- Resource conservation
  - Installation of second reverse osmosis system in the power plant has resulted in the saving of 196 kl/day of fresh water.
  - Installation of air-cooled fluid cooler in the power plant has led to the saving of 720 kl/day of water.
  - The STP generates 350 kl of treated water per day, which is used for watering plants and controlling dust in the plant facilities.
  - The project has resulted in the saving of 720 kl/day of water and approximately 14 040 tonnes/annum of fuel.

- *Minimization of waste* Steam generation from waste heat of pre-heater gases has eliminated the problem of ash generation.
- *Greenhouse gas mitigation* The project has succeeded in reducing the emissions of 76 000 tonnes of CO_2 per annum due to the burning of fuel.

CASE STUDY AT A GLANCE

- Shree Cement Ltd, with its present capacity of 6.8 MTPA, is the largest plant at a single location in North India.
- The company also has a captive power generation unit of 101.5 MW capacity.
- The company's annual turnover in 2007/08 was Rs 20.66 billion, and it is all set to touch Rs 300 billion mark in 2008/09.
- The company has taken various measures for conservation of natural resources like water and fuel.
- The company has initiated several water conservation measures by installing second reverse osmosis system and air-cooled fluid cooler at the power plant, and by recycling of domestic waste water using a sewage treatment plant.
- In its cement plant, the company has installed a waste heat recovery boiler to use waste heat to generate fuel, in the process saving 720 kl/day of water and approximately 14 040 tonnes/annum of fuel.
- Net savings accrued by installing the waste heat recovery boiler are Rs 95.2 million per annum.

CASE STUDY 17

Bharat Aluminium Company Ltd

Figure 1
Bharat Aluminium Company Ltd

About the company

Bharat Aluminium Company Ltd, or BALCO, located at Korba in Chhattisgarh, is the leading aluminium manufacturing company in India (Figure 1). It was incorporated in 1965 as a public sector undertaking. BALCO has several 'firsts' to its credit. It is the first public sector enterprise in the country that started producing aluminium in 1975. In 1987/88, a captive power plant of 270-MW capacity was added to cater to the power requirement of the unit. BALCO has been the first in the Indian aluminium industry to produce alloy rods, which is a feedstock for all aluminium alloy conductors needed for today's power transmission lines. BALCO is part of Vedanta Resources, a London-listed metals and mining major, with aluminium, copper, and zinc operations in the UK, India, and Australia.

In the last 41 years, BALCO has developed alumina production capacity of 200 000 tonnes per annum and smelting capacity of 350 000 tonnes per annum, and has expanded its fabrication facility to include three Properzi rod mills, three pig casting machines, integrated hot and cold rolling mills, and captive power plants of 810-MW capacity.

The turnover of the company in 2007/08 was Rs 46 105.4 million.

Environmental profile

BALCO has a combined Quality, Environment, and Occupational Health and Safety Policy. The policy has the following environmental components.

- Conserving natural resources like raw materials and water, and energy by process improvement, recycling, and reducing waste.
- Preventing pollution and complying with all environmental rules and regulations in all its operations.

The company also has an environment department and an Environment Management System in place. The company received ISO 9001:2000 for manufacture and supply of calcined alumina, aluminium ingots/billets, aluminium rods, and aluminium rolled products.

Air pollution

The main air pollutants emitted by the facility are SPM (suspended particulate matter) and fluorine gases.

Air pollution control measures

To control air pollution, various cleaner technologies have been implemented and state-of-the-art pollution control equipment have been installed, as listed below.

- Centre worked pre-baked technology
- Dry scrubbing system
- High efficiency ESP (electrostatic precipitator)
- Bag filters
- Battery-operated vehicles in pot rooms
- Dry ash evacuation system
- 275-metre-high stacks for effective disposal of pollutants
- Water fogging in the ESP inlet by using special nozzles
- Celtrol modification
- Wet drilling
- Black topping of roads

Water pollution

pH and oil and grease are the major pollutants found in the waste water generated from the facility.

Water pollution control measures

pH

The effluent generated from the alumina plant usually has a high pH. To treat this alkaline effluent, a neutralization system has been provided in the plant. The neutralization system comprises an automatic acid dosing system and three online pH meters. In the neutralization system, acid is added to the effluent to bring down its pH level to 5.5–9. About 80% of the treated effluent is recycled back into the system.

Oil and grease

The waste water generated from the fabrication plant contains oil and grease. This water is treated in the oil water separation system and emulsion treatment plant (Figure 2).

Figure 2
Emulsion treatment plant

Hazardous waste

The various types of hazardous waste generated by the facility are cathode residues, including pot lining waste, sludge from the gas treatment plant, used/spent oil, sludge from the emulsion treatment plant, used lead-acid batteries, and waste containing oil.

Hazardous waste management

BALCO has an efficient hazardous waste management system in place. The initiatives taken by the company to dispose of the hazardous waste are as follows.

- *Cathode residues, including pot lining waste* The spent pot lining waste is reused in cement and steel industries, as per the recommendations of the CPCB (Central Pollution Control Board). The fluoride salt recovered from this waste is used in BALCO smelters. The company is also in the process of utilizing the spent pot lining waste as a fuel in its power plants, along with coal. The design of the pot has been modified, which has resulted in the reduction in pot lining waste generation.
- *Sludge from the gas treatment plant* This waste contains fluorine in a large quantity, which is being recovered on a pilot scale.
- *Used/spent oil* This oil is sold to the MoEF (Ministry of Environment and Forests)–authorized oil recyclers.
- *Sludge from the emulsion treatment plant* This waste is incinerated in an incinerator.
- *Used lead-acid batteries* These are disposed of to the MoEF-registered recyclers.
- *Waste containing oil* This waste is generated from all maintenance departments and is incinerated in boilers/furnaces or in the incinerator.

Other waste

Red mud

About 950 MT of red mud is generated each day during the production of alumina from bauxite. This red mud is disposed of using high concentration slurry disposal system instead of wet disposal method. So far, there is no high volume utilization of this waste.

Fly ash

Fly ash generated from boilers in the steam plant is used for the reclamation of low-lying areas and in the brick making industry. The fly ash generated from captive boilers is also used for manufacturing cement and for raising ash dyke.

Case study

Replacement of the wet disposal system with high concentration slurry disposal system for the disposal of red mud

Generation of red mud

Alumina is produced in an intermediate stage during the process of manufacturing aluminium from bauxite. Bauxite is crushed and wet ground in a ball mill with spent liquor (caustic soda) from the process to make slurry. The slurry is preheated in low-pressure heaters and subjected to desilication to remove silica from the slurry. After removing silica, the slurry is again heated in high-pressure preheaters and pumped into a series of digesters operating at high temperature and pressure. The bauxite dissolves, and sodium aluminate and oxides of soda, aluminium, and silica are formed. The alumina formed from bauxite is chemically transferred into a liquor known as pregnant liquor, or green liquor. The pressure and temperature of the slurry are subsequently brought down by passing it through a series of flash tanks in which steam is used for preheating the ground slurry, which is then sent to digestion tanks. The slurry from the last digestion tank is finally pumped to settlers to separate the red mud.

The settled mud is subsequently washed in a series of washers to recover caustic soda. The recovered caustic soda is pumped to the settlers, where it joins the main stream containing green liquor.

Wet disposal method

In the wet disposal method followed earlier, the washed mud, a mixture of 20% solid and 80% water, was filtered and disposed of to a red mud pond. This mud settled at the bottom of the pond. The supernatant liquor from the red mud pond was recycled back into the process in the alumina refinery. Moreover, the average life of the pond was five years, resulting in frequent acquisition of land for pond construction, a cost-intensive activity.

High concentration slurry disposal system

To address the drawbacks of the wet disposal method, an environment-friendly disposal method, that is, high concentration slurry disposal system, was introduced (Figure 3).

Figure 3 High concentration slurry disposal system

In this method, red mud filter cakes are pumped to a shearing tank, where the red mud is conditioned to the correct rheological behaviour. This ensures that the mud spreads out uniformly on the disposal site. The reconditioned mud is transferred to a piston diaphragm pump with the help of a suitable feed pump. The density of the red mud is monitored constantly throughout the process. From the piston diaphragm pump, the red mud is transferred to the red mud pond.

This method has resulted in the reduction in the overall water consumption and increased the capacity utilization of red mud ponds. The solid content of the red mud separated using this process is higher than 50%, which is as per the norm of the CREP (Corporate Responsibility for Environmental Protection).

Cost–benefit analysis

Parameter	*Wet red mud disposal system*	*High concentration slurry disposal system*
Pond cost	Rs 220 million	Rs 220 million
Life of pond	5 years	12.5 years
Life of abandoned ponds	0	7.5 years
Cost of pond/year	Rs 44 million	Rs 11 million
Power cost/year	Rs 2.93 million	Rs 2.93 million
Water cost/year	Rs 365 000	0
Caustic soda saving/year	0	Rs 2.7 million
Total cost/year	Rs 47.29 million	Rs 13.93 million

Savings envisaged per year are to the tune of Rs 33.36 million, and the payback period is 5.4 years.

Benefits

- *Conservation of water* After introducing high concentration slurry disposal system, overall water consumption has reduced to about 500 m^3/day. In the wet disposal method, the red mud comprised 80% water and 20% solids, while in the high concentration slurry disposal system, the red med comprises 55% solids and 45% water.
- *Conservation of land* During the wet disposal of the red mud, the red mud pond used to get filled in five years. But in the high concentration slurry disposal system, a red pond mud can be utilized for 12.5 years. This has led to significant land conservation.
- *Recovery of caustic soda* About 214 tonnes of caustic soda is recovered every year from the high concentration slurry disposal system. The net savings due to this recovery are Rs 2.7 million/year.

Case study

Replacement of the wet scrubbing technology with the dry scrubbing technology

Generation of fluorides

The process of production of aluminium by electrolysis of alumina in a molten cryolite bath at a temperature of 960 °C emits fluoride in the form of gaseous HF (hydrogen fluoride) and particulate fluorides comprising sodium aluminium tetrafluoride, cryolite, and AlF_3 (aluminium fluoride). The fumes containing fluorides are treated in the fume treatment plant (Figure 4).

Figure 4
Fume treatment plant

Wet scrubbing process

The wet scrubbing process was developed in the 1950s to control fluoride emissions, but its efficacy to capture submicron particles was limited. Also, the conversion process of the captured fluorides to produce cryolite or AlF_3 was complex and expensive.

The wet scrubbing process employed by BALCO had two stages. The first stage included the removal of coarse particles from the fume by using a cyclone. These particles were recycled back into the process by blending them with the primary alumina. This process was not very efficient. The second stage consisted of a series of water sprays. In this stage, the fume used to come in contact with water spiked with alkaline chemicals, and in the process, HF was absorbed by water. This water containing HF was then pumped to the cryolite regeneration plant, where fluoride was removed by precipitating cryolite with the help of expensive chemicals. The precipitated cryolite was dried to a solid in a rotary kiln and recycled back into the smelting process.

The wet scrubbing process consumed a large quantity of water, and it did not scrub all the HF and particulates. Besides, it used very expensive chemicals.

Dry scrubbing process

In 1958, Pechiney showed that the reactive alumina has the potential to absorb gaseous fluoride. Gaseous fluoride reacts with alumina to form a stable compound that does not desorb when fluorinated alumina is fed to electrolysis pots.

These advantages of using alumina made BALCO to phase out the wet scrubbing technology and replace it with the dry scrubbing technology. It works on the simple premise that HF in the gaseous form comes in contact with alumina in a reactor. The alumina absorbs the fluoride and returns fluoride-rich alumina to a silo, ready for recycling into the smelting process.

The dry scrubbing process essentially comprises an injection-type scrubber, which uses a simple vertical venture reactor to establish contact between the gravity-injected alumina and the upflow process gases. There is also a fabric filter, which collects the alumina and particulate fluorides, acting as a secondary reactor to complete the adsorption of HF via the alumina coating on the bags. To ensure contact between the alumina and the gaseous HF, the alumina is recycled 10–20 times into the venture reactor, which increases the alumina concentration in the reactor. A fluidized bottom in the filter hopper provides the required alumina storage to continuously recycle alumina to the reactor, while a simple overflow chute discharges fluorinated alumina onto an air slide and back to pots.

Thus, dry scrubbers work in a closed loop fashion. The fluorides emitted at the pots are captured by dry scrubbers and returned to the pots, without any intermediate process or treatment.

Cost–benefit analysis

Total cost of the installation of the dry scrubber: Rs 379 million
Savings due to the reduced consumption of AlF_3: Rs 432/MT of aluminium produced
Savings due to the reduced consumption of cryolite: Rs 97.8/MT of aluminium produced
Total expected quantity of aluminium produced in a year: 115 000 MT
Total expected annual savings due to the installation of dry scrubber: Rs 60.9 million
Payback period: 6.2 years

Benefits

- More than 95% reduction in fluoride emissions has been achieved. Fluoride emissions from the stacks of the facility used to be in the range 7–8 mg/Nm^3 when the wet scrubbing process was in place. After implementing the dry scrubbing process, the emissions from the dry scrubbers have reduced to 0.3–0.5 mg/Nm^3.
- Phasing out of the wet scrubbing process has reduced the water consumption by 70 m^3/day, as no water is used in the dry scrubbing process.
- The wet scrubbing process used to generate black sludge, a hazardous waste. No such sludge is generated in the dry scrubbing process.
- Particulate emissions in the wet scrubbing process were in the range 50–90 mg/Nm^3. These emission levels came down to 20 mg/Nm^3 in the dry scrubbing process.
- There has been substantial reduction in the raw material (AlF_3 and cryolite) usage.

Case study

Online cutout of GP320 aluminium reduction cell at 320 kA

GAMI technology

BALCO uses GP320 reduction cells for producing aluminium using GAMI technology. Aluminium is produced through electrolysis process (Figure 5). Electrolysis takes place in a cell called pot, or reduction cell, which consists of rodded carbon anodes, electrolyte called bath, and carbon cathodes as the bed of the pot. A current of 320 kA is passed through these pots connected in series for the electrolysis.

In GAMI technology (GP320), pots are cut-off from the pot line circuit by taking complete power outages. To take a pot off the line, the pot line amperage is reduced to 0 kA. This whole process of cutting out pot takes about 30–35 minutes, with 15–20 minutes at zero line current. When the pot line current is reduced to zero, the resistance of the pot increases, as the bath

Figure 5
Balco Gami GP-320 potline

electrolyte temperature in the pot starts decreasing. Thus, the average voltage of the pot also starts increasing. And when the voltage becomes more than 8 volts, the cell goes on anode effect, causing the release of gases like CF_4 and C_2F_6.

The power outages have the following disadvantages.

- Higher anode frequency, which results in higher CF_4/C_2F_6 emissions.
- Disturbances to the operating pots.
- Impact on the operations and equipment of the power plant.
- Production losses due to a decrease in line amperage as well as zero amperage during cut out of pots.

Modifications in the technology

BALCO has developed a procedure to cut-off pot online, that is, without reducing the line amperage to 0 kA. Prior to the cut-out of the pot, necessary preparations in the pot have to be done. The anode crust has to be broken at the four corners of the pot. Then, about 5–6 tonnes of liquid bath has to be tapped out. While tapping out liquid bath, anodes are dipped into the metal pad, bringing down the pot voltage to 1.5 V. Once the pot voltage is reduced, cut out operation is carried out in normal procedure. In this case, there is no need for lowering down the line amperage to 0 kA, as the whole process is carried out at 320 kA. BALCO has successfully implemented this technology in the cut-out of 50 pots and adopted this as a standard procedure to put pots off power.

Table 1 Energy lost because of load throw off during power outage taken for pot cut off

Unit no.	Stable load before load throw off (MW)	Duration at 0 kA (minutes)	Minimum load reached during throw off (MW)	Stable load after load throw after stabilization (MW)	Million Units (MU) lost
1	120.1	20	68.27	124.13	0.0186
2	122.5	20	69.74	132.6	0.0210
3	125	20	63.3	129	0.0219
4	114.9	20	65.1	126.57	0.0205
Station	482.5		266.4	512.3	0.0820

Cost–benefit analysis

The cost incurred by this project is very nominal, as it requires only labour cost for less than 3 minutes to remove the insulation plates from the 5 risers of each pot. So, the investment is negligible.

Benefits

- Implementation of this project has resulted in less anode effect frequency and thus less PFC (CF_4 and C_2F_6) emissions.
- A total of 0.082 MU is lost for duration of 20 minutes at 0 kA because of load throw-off during power outage taken for pot cut off (Table 1). By cutting out pots online at 320 kA, load throw off is eliminated and power is utilized for production.
- Production losses due to the non-utilization of power thrown off due to the lowering of pot line amperage to 0 kA for cutting out pot is around 9.6 MT of metal. The online cutting of pots has eliminated this production loss.
- Disturbances to the operating pots due to frequent power outages are eliminated.
- There is no negative impact on the power plant operation as well as on its equipment.

CASE STUDY AT A GLANCE

- Bharat Aluminium Company Ltd, or BALCO, located at Korba in Chhattisgarh, is the leading aluminium manufacturing company in India.
- The turnover of the company in 2007/08 was Rs 46 105.4 million.
- The company has replaced the wet disposal system with high concentration slurry disposal system for the disposal of red mud.
  - This method resulted in the reduction in water consumption and increased the capacity utilization of red mud ponds. The solid content of the red mud separated using this process is higher than 50%, which is as per the norm of the CREP (Corporate Responsibility for Environmental Protection). This process also led to land conservation.
  - Savings envisaged per year are to the tune of Rs 33.36 million.
- The company has replaced the wet scrubbing technology with the dry scrubbing technology for reducing fluoride emissions.
  - More than 95% reduction in fluoride emissions has been achieved.
  - The water consumption has reduced by 70 m^3/day, as no water is used in the dry scrubbing process.
  - No sludge is generated in the dry scrubbing process.
  - There is a substantial reduction in raw material (AlF_3 and cryolite) usage.
  - Total expected annual savings due to the installation of dry scrubber are Rs 60.9 million.
- The company uses GP320 reduction cells for producing aluminium using GAMI technology.
- In GAMI technology (GP320), pots are cut off from the pot line circuit by taking complete power outages and the pot line amperage is reduced to 0 kA.
- The company has developed a procedure to cut-off pot online, that is, without reducing the line amperage to 0 kA.
- The company has successfully implemented this technology in the cut-out of 50 pots and adapted this as a standard procedure to put pots off power.
- This process has resulted in the following benefits.
  - Implementation of this project has resulted in less anode effect frequency and thus less PFC (CF_4 and C_2F_6) emissions.
  - A total of 0.082 MU is lost for duration of 20 minutes at 0 kA because of load throw off during power outage taken for pot cut off. By cutting out pots online at 320 kA, load throw off is eliminated and power is utilized for production.
  - Production losses due to the non-utilization of power thrown off due to the lowering of pot line amperage to 0 kA for cutting out pot is around 9.6 MT of metal. The online cutting of pots has eliminated this production loss.
  - Disturbances to the operating pots due to frequent power outages are eliminated.
  - There is no negative impact on the power plant operation as well as on its equipment.

CASE STUDY 18

Mangalore Refinery and Petrochemicals Ltd

Figure 1
Mangalore Refinery and Petrochemicals Ltd

About the company

The MRPL (Mangalore Refinery and Petrochemicals Ltd) is a grass-root refinery, set up as a joint venture company by the Hindustan Petroleum Corporation Ltd and Indian Ryon Industries Ltd. Phase I Refinery with 3.69-MTPA (million tonnes per annum) capacity was commissioned in 1996 and Phase II Refinery was commissioned in 1999. The ONGC (Oil and Natural Gas Corporation) took over the MRPL in 2003, and presently, the MRPL is a subsidiary company of the ONGC.

The Refinery is located on a beautiful hilly site, approximately 20 km north of Mangalore city and 350 km away from Bangalore city, in the state of Karnataka (Figure 1). Being a petroleum refinery, the MRPL produces liquefied petroleum gas, naphtha, petrol, kerosene, aviation turbine fuel, diesel, furnace oil, bitumen, mixed xylene, and sulphur by-product. The MRPL's acquisition by the ONGC helped in the downward integration of the oil major ONGC in India. During the financial year 2007/08, the Refinery processed a record throughput of 12.55 MT (million tonnes) of crude oil, achieving 130% capacity utilization. It dispatched 11.83 MT of finished products. The company has achieved a turnover of Rs 373 390 million in 2007/08, registering a net profit of Rs 12 720 million.

Environmental profile

An environment-conscious company with environment-friendly operations, the MRPL has an Environment Policy in place, which is followed diligently. The company is ISO 14001/9001 certified and its EMS (Environment Management System) strictly follows system standards. The company regularly conducts internal audits to improve its EMS.

Air pollution

The main air pollutants emitted from the Refinery are SO_2 (sulphur dioxide), NO_x (oxides of nitrogen), hydrocarbons, and H_2S (hydrogen sulphide).

Air pollution control measures

The various measures taken to mitigate the air pollutants are as follows.

- Less than 1% (average) sulphur is maintained in the furnace oil burnt in all the heaters/furnaces. This reduces SO_2 emissions from the Refinery.
- Fuel gas generated in the Refinery is sweetened to less than 100 PPM (parts per million) sulphur and completely utilized in the Refinery.
- Tall stacks are provided for all heaters (higher than the CPCB [Central Pollution Control Board]/KSPCB [Karnataka State Pollution Control Board] stipulation), which ensures effective dispersion and dilution of various air pollutants.
- Facilities like amine treatment, sour water stripping, and sulphur recovery units improve air quality by minimizing the emissions of SO_2 and help in recovering elemental sulphur.
- Low-NO_x burners are provided in all furnaces, which effectively reduce NO_x emissions from the stacks.
- Hydrogen emissions are minimized by using floating roof tanks for storing hydrogen, crude, and products.
- Efficient flaring system is in place to completely burn the waste gases from the vents of process units.
- Sulphur recovery units have been installed in the facility.
- Merox treatment plants have been set up for reducing H_2S and removing mercaptans.
- Efficient heat recovery system is employed for reducing emissions.

Table 1 gives the observed values of air pollutants against the prescribed standards.

Water pollution

Waste water generated in the Refinery comprises water pollutants like sulphides, oil and grease, phenols, suspended solids, COD (chemical oxygen demand), BOD (biochemical oxygen demand), and ammoniacal nitrogen.

Water pollution control measures

State-of-the-art waste water treatment plants are in place in the Refinery. These plants operate round-the-clock. There are online analysers for measuring pH, sulphides, phenols, dissolved oxygen, and total organic carbon. The waste water treatment plant comprises the following sections.

Table 1 Emission of air pollutants vis-à-vis standards

Pollutant	Technology/system adopted	Emission: Prescribed standard (mg/m^3)	Emission: Observed value (mg/m^3)	Percentage violation
SO_2	Using high volume sampler as NAAQS	80	16.93	Nil
NO_x	Using high volume sampler as NAAQS	80	1.75	Nil
H_{2S}	Using high volume sampler as NAAQS	–	0.01	Nil
CO	Using high volume sampler as NAAQS	5 mg/Nm^3	<1.0 mg/Nm^3	Nil
HC	Using high volume sampler as NAAQS	–	0 PPM	Nil
SPM	Using high volume sampler as NAAQS	120	26.37	Nil

SO_2 – sulphur dioxide; NO_x – oxides of nitrogen; H_2S – hydrogen sulphide; CO – carbon monoxide; HC – hydrocarbon; SPM – suspended particulate matter; NAAQS – National Ambient Air Quality Standards; PPM – parts per million

- Section I: Free oil removal section followed by chemical treatment section for the removal of hydrocarbons, sulphides, phenols, COD (partially), and emulsified oil.
- Section II: Biological treatment section for the removal of BOD, COD, sulphides, phenols, nitrogen, and nitrogen compounds.
- Section III: Tertiary treatment section employing filtration through sand and activated carbon for the removal of suspended solids (polishing system) and organics.
- Section IV: Slop oil handling system and sludge dewatering and disposal system.
- Section V: Chemical dosing system.

About 70%–75% of the treated effluent is recycled/reused in cooling towers on a continuous basis. Table 2 gives the standards achieved in the Refinery against the limits stipulated by the KSPCB and MINAS (Minimum National Standards).

Hazardous waste

The hazardous waste generated by the facility and its disposal mechanism are mentioned as follows.

- Oil and oily sludge generated from the crude tank bottoms are treated with hot gas oil (soil extraction method) for maximum recovery of hydrocarbons.
- Further, enhanced recovery of hydrocarbons is achieved by treating the sludge with chemicals. Final treated sludge comprising mainly silicates, traces of heavy metals, and

Table 2 Standards achieved in the treated effluent

Parameter	KSPCB/MINAS standards	Standards achieved in the Refinery
pH	6.0–8.5	Always around 7.0
Oil and grease	10 mg/l	3–5 mg/l
Suspended solids	20 mg/l	<15 mg/l
Sulphides	0.5 mg/l	Non-detectable
Phenol	1.0 mg/l	<0.3 mg/l
COD	250 mg/l	<100 mg/l
BOD	15 mg/l	<7 mg/l
Ammoniacal nitrogen	50 mg/l	<10 mg/l

KSPCB – Karnataka State Pollution Control Board; MINAS – Minimum National Standards; COD – chemical oxygen demand; BOD – biochemical oxygen demand

hydrocarbons is bioremediated using TERI technology and finally landfilled in secured landfills or disposed of using land farming techniques.

- Oily sludge generated in waste water treatment plants is treated in sludge thickeners and centrifuges, and the oil-recovered sludge with 3%–5% oil is taken up for bioremediation.
- Spent lube oil/spent transformer oil generated in the Refinery is re-processed by mixing the same in the crude oil.
- Spent batteries are given away to the ministry or KSPCB-authorized recyclers or buyers.
- Mercury from tube lights is recovered and sent to vendors for reuse.

Case study

Energy conservation and pollution reduction projects

Various energy conservation projects were undertaken in the Refinery with a view to save energy. The pollution reduction projects undertaken by the company managed to significantly reduce atmospheric emissions from the facility by way of reducing fuel burnt in various heaters across the Refinery, which enhanced the profits of the company. These projects are described briefly in the following sections.

Making Euro-3 and Euro-4 motorspirit

The MRPL had the technology and expertise to make BS-2 grade MS (motor spirit), but did not have the proper infrastructure to make Euro-3 and Euro-4 MS till 2005. To achieve required aromatic specifications of Euro-3 and Euro-4, a neutral octane blending stream is necessary. An ISOM (isomerization) Unit of 2100 million tonnes per day capacity was commissioned in 2006/07 to reduce benzene and sulphur contents in the products, at a cost of Rs 2330 million.

The ISOM unit has the following sections.

- Light naphtha hydrotreating section
- Benzene saturation section
- Penex section

The ISOM unit gives products with following qualities.

- High octane (preferably ≥91)
- Zero benzene
- Zero aromatics
- Low density stream

Table 3 gives a comparison between BS-2, Euro-3, and Euro-4 MS.

Table 3 Comparison between BS-2, Euro-3, and Euro-4 motor spirit

Parameter	BS-2	Euro-3	Euro-4
Specific gravity	0.71–0.77	0.72–0.775	0.72–0.775
Sulphur (PPM)	500	150	50
RON (research octane number)	89	91	91
Benzene (vol%)	5/3	1	1
Olefin (vol%)	No specification	21	18
Aromatics (vol%)	No specification	42	35

Table 4 gives the details of emission reductions achieved.

Table 4 Emissions reduction achieved

Year	Sulphur dioxide reduction (tonnes/year)	Benzene reduction (M^3 per year)
2006/07	88	3280
2007/08	411	15 260

Implementing variable speed drives for 3 MV and 119 LV drives

In general, all motors run at a constant and rated speed. As the process requires variable flow requirement, depending on annual/monthly targets, the same is achieved by throttling the control valve. So, unintentionally, electrical power is fully utilized in motor, and the excess energy is lost as pressure drop in the control valve. The MRPL Encon and the environment team together identified potential energy lost in process units. By introducing variable speed drive mechanism in identified motors, the estimated power savings were to the tune of 2.7 MWh. Thus, the MRPL has implemented variable frequency drives for as many as 119 LV (low voltage) drives and 3 MV (medium voltage) drives to optimize power consumption for individual motors in the Refinery. After implementing variable speed drives in various motors, the fuel consumption also reduced in furnaces/heaters.

Cost–benefit analysis

Power saving to the tune of 2.7 MWh will reduce furnace oil usage in various heaters by 6700 MT per year. The project was implemented in 2005 with an investment of Rs 200 million. Net savings accrued are Rs 88 million and payback time is 2.3 years. SO_2 emission reduction achieved is 134 MT per year. Table 5 summarizes the company's achievement in terms of pollution reduction and energy saving.

Table 5 Pollution reduction and energy saving

Project description	*Achievements in energy savings on per year basis*				*Emission reduction (MT/year)*		
	Ele (million kWh)	*Fuels–fuel oil (kl)*	*Fuel (Mkcal)*	*Total (Rs million)*	*Oxides of sulphur*	*Oxides of nitrogen*	*Carbon dioxide*
Installing 3 medium and 119 low voltage variable speed drives	21.6	6688	64 890	88.26	127	13.52	20 600

Utilization of waste in biogas plant

The MRPL commissioned an environment-friendly biogas generation plant in August 2007 for generating biogas from canteen and colony food waste. The plant comprises feed shredding facilities, primary and secondary anaerobic digesters, a biogas collection system, a biogas compression system, and a delivery system to canteen burners. The plant can handle up to 1000 kg/day of food and vegetable waste. The capacity was fixed based on estimates on food waste generated per day per household, current canteen waste, and future colony/canteen expansion.

The compressed biogas is supplied to canteen, where it is used as a fuel substitute for commercial liquefied petroleum gas.

Process description

The biogas plant operates on the UASB (upward anaerobic sludge blanket) process. The food waste is sorted to remove non-biodegradable materials like plastic, aluminium foil, and so on. It is then shredded in a shredder. It gets converted into slurry, which is then introduced into a modified UASB primary digester. The digester mainly serves as a hydrolysis-cum-acidification tank for treating suspended solids. The overflow of the primary digester is collected in a sump-cum-recycle chamber and pumped into the secondary digester.

The secondary digester serves as the main methane fermentation tank, where BOD reduction takes place. The treated overflow from this digester is drained. The sludge, which accumulates at the bottom of the digester, is removed periodically.

The biogas generated is collected in a neoprene rubber balloon. A compressor is used for compressing the gas, which is then supplied to the MRPL canteen. On a typical feed quantity of 1000 kg/day, the plant can generate about 70–80 M^3 of biogas.

Amine flash column off-gas to SRU-2 incinerator

An in-house modification was done to recover heat from flash column off-gases, which were earlier directly sent to flare. These gases mainly consist of lighter hydrocarbons. With this modification, the amine flash column off-gases are now routed to the SRU-2 (sulphur recovery unit) incinerator. In the incinerator, these gases are burnt off, and heat is recovered.

An off gas KOD (knock-out drum) was put in the line at the SRU-2 to knock off any trace of liquid for the purpose of safety. Another line to the SRU-3 incinerator, along with KOD, is being planned to utilize the flash gas heat during the shutdown of the SRU-2 unit (Figure 2).

Figure 2 Sulphur recovery unit-2 incinerator

Cost–benefit analysis

Total investment made for this modification is Rs 1.02 million. Thermal energy savings are to the tune of 969 SRFT (standard refinery fuel tonnage)/year, or Rs 13.57 million/year. Payback period is 27 days. SO_2 emission reduction achieved is 19.186 MT/year.

Optimization of SWS operation

A study was carried out to check the feasibility of optimizing Refinery SWS (sour water stripper) operation. Earlier, both Refinery SWS-1 and Refinery SWS-2 units were required to run because of reboiler limitations due to tube leak. Once it was checked, SWS-2 unit capacity was sufficient to take care of both ph1 and 2 refinery SWS loads.

Accordingly, SWS-1 load was shifted to SWS-2 and operations in the SWS-1 unit was stopped. This resulted in the saving of reboiler steam and power.

Cost–benefit analysis

There was no investment. Thermal savings are to the tune o 1965 SRFT/year, or Rs 28 million/year and electrical savings are of the order of 470 MWh/year.

Antifouling injection in CDU-1 feed

In the CDU-1 (crude distillation unit), where a large amount of crude from Mumbai High is processed, fouling rate of preheat exchangers was high. To minimize this fouling rate, antifoulant injection was started on a trial basis in 2006, and this experiment was found to be successful.

Cost–benefit analysis

The investment made was negligible. Thermal energy savings are to the tune of 1277 SRFT/year, or Rs 18 million /year. Chemical cost accrued is Rs 6 million/year. Net savings in 2006/07 were Rs 12 million/year. SO_2 emission reduction achieved is 25.28 MT.

VBU-2 fuel oil heat recovery by CDU-2 crude preheating

In the VBU-2 (visbreaking unit), fuel oil product was being cooled using tempered water system. The system comprises tempered water circulation pump and fin fans. Cold crude feed from OM&S to the CDU-2 pipeline was transferred through an interconnecting pipe rack. This crude was initially preheated with LP (low pressure) steam and finally with fired heater, before it entered the atmospheric column.

The system was studied for utilizing waste heat from VBU-2 fuel oil to preheat the CDU-2 crude. Since the crude pipeline is located nearby, the length of the crude pipeline added was small. The modification was carried out by cutting the crude line at the northwest corner of the VBU-2 and routing it to the VBU-2 via manifold. Near the tempered water cooler, the crude line was connected to tempered water line. After commissioning the system, tempered water system in the VBU-2 has been stopped, and LP steam consumption in the CDU-2 has come down (Figure 3).

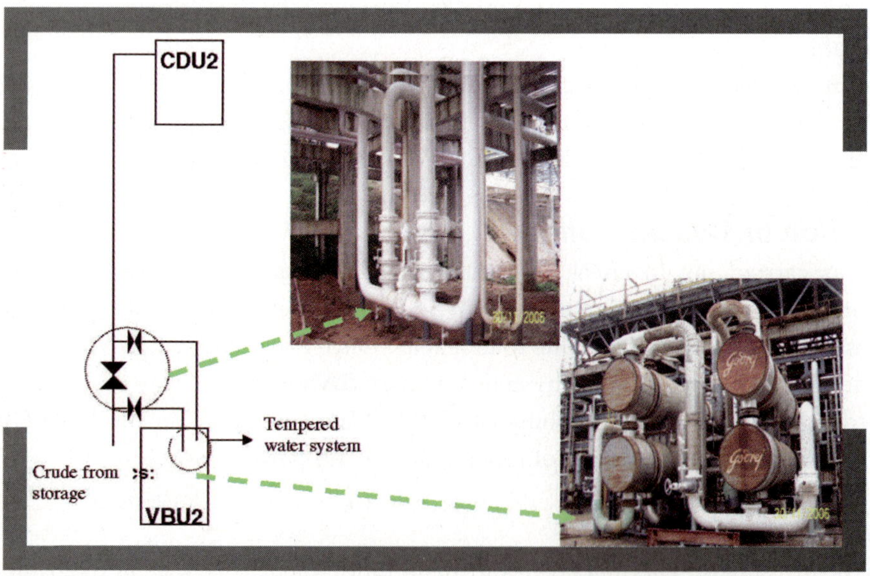

Figure 3 Crude distillation unit and visbreaking unit: fuel oil heat recovery

Cost–benefit analysis

Thermal energy savings at the CDU-2 are to the tune of 10.7 MT/h LP steam, or 5006 SRFT/year, or Rs 70.08 million/year. Electrical energy savings at VBU-2 are 200 kW power, or 1670 MWh/year, or Rs 6.48 million/year. Total savings accrued are 16 SRFT/day, or 5469 SRFT/year, or Rs 76.56 million/year. The total investment made was Rs 3.1 million and payback period is 15 days. SO_2 emission reduction achieved is 108.29 MT.

Provision of secondary seals to 16 storage tanks to reduce fugitive emissions

Ten existing and six new storage tanks (all floating roof) were provided with secondary seals to minimize fugitive emissions.

Heater decoking by hydrokinetics

Earlier, decoking was being carried out in a traditional method of burning coke with air. This process was time consuming and energy intensive. Hydrokinetics technology was introduced in which decoking was carried out without the burning of coke. Instead, it was removed by means of bullets. This eliminated the usage of utilities and saved time.

Cost–benefit analysis

Thermal energy consumption in normal decoking is Rs 3.5 million/decoking, while no thermal energy is consumed in decoking by hydrokinetics. The project cost is Rs 2.75 million/decoking. Net savings accrued are Rs 0.75 million/decoking. Net savings accrued in 2006/07 were Rs 1.5 million/year.

Optimization of over flash at hydrocracker units 1 and 2

In hydrocracker units 1 and 2, over flash flow was higher than design. This led to higher heat duty of non-UOP heater and higher pumping energy requirement. The over flash was also utilized as reboiling media in kerosene stripper. But approximately, 30 m^3/h of over flash was only used for the reboiler heat duty, and the remaining was passed through the bypass. The saving potential was in minimizing this bypass flow. Accordingly, the flow was reduced from 75/70 m^3/h to 65/65 m^3/h.

Cost–benefit analysis

Thermal energy savings at hydrocracker units 1/2 are 0.18/0.05 MMkcal/h, or 184 SRFT/year, or Rs 2.58 million/year. SO_2 emission reduction achieved is 3.64 MT/year.

Changing of pressure swing adsorption adsorbents in hydrogen-2 and continuous catalytic reformer-2 units to improve hydrogen recovery

PSA (pressure swing adsorption) adsorbents were changed and value refurbishing was done in hydrogen-2 and CCR (continuous catalytic reformer)-2 units to improve the PSA hydrogen recovery. This helped in reducing the hydrogen loss.

Cost–benefit analysis

Estimated savings due to the increased hydrogen recovery are 2375 Nm^3/h, which is equivalent to 8000 MT of LSHS (low sulphur heavy stock)/year, or Rs 115 million/year. The investment made was Rs 110 million and the payback time is one year. SO_2 emission reduction achieved is 160 MT/year.

Heat recovery from splitter bottoms for feed preheating at the hydrocracker unit-1

The intermittent bleed from hydrocracker splitter bottoms to LSHS was shifted from upstream to downstream of 'splitter bottom vs splitter feed' heat exchanger, which aided in the recovery of heat energy, which was being cooled in fin-fan cooler.

Cost–benefit analysis

Estimated savings due to waste heat recovery are 130 MT of LSHS/year, or Rs 1.87 million/year. The investment made was Rs 0.2 million and the payback period is two months. SO_2 emission reduction achieved is 2.6 MT/year.

Stabilizer reboiler desuperheater at hydrocracker unit-1

At the hydrocracker unit-1, stabilizer reboiler media was MP (medium pressure) steam, which was at superheated condition at 280 °C. Hence, the efficiency of the reboiler load was less. A modification was done by adding a de-super heater at the upstream of reboiler. There was a net saving in the MP steam consumption, along with an increase in yield.

Cost–benefit analysis

Fuel gas reduction was of the order of 0.11 MT/h of MP steam, which is equivalent to 58 SRFT/year, or Rs 0.74 million/year.

Heat recovery from product diesel for feed preheating at the hydrocracker unit-2

A modification was made to route CDU-1 VGO (vacuum oil gas) to the hydrocracker unit-2 as part of optimization. This VGO line was routed to downstream of 'HSD (high speed diesel) vs VGO' exchanger at the hydrocracker unit-2 to maximize waste heat recovery.

Cost–benefit analysis

Estimated savings due to waste heat recovery are 300 MT of LSHS/year, which is equivalent to savings worth Rs 4.3 million/year. Investment made was Rs 0.15 million and the payback period is one month. SO_2 emission reduction achieved is 6.0 MT/year.

Interconnection of hydrocracker 1 and 2 stabilizers

The stabilizer feed sections of the hydrocracker units were interconnected using the HPNA (heavy polynuclear aromatics) nitrogen recirculation line to avoid flaring of stabilizer feed gases during maintenance shutdown of stabilizer feed gas compressors. This interconnection reduced light hydrocarbon (including LPG) flaring and minimized hydrocracker throughput reduction during the maintenance of stabilizer compressor.

Cost–benefit analysis

Net savings due to waste heat recovery are 400 MT of LSHS/year, which is equivalent to Rs 5.7 million/year. Investment made is Rs 0.5 million and payback period is two months. SO_2 emission reduction achieved is 8.0 MT/year.

Routing of hydrogen-rich hydrocracker reactor low-pressure separator off gas to gas oil hydrodesulphurizer

Hydrocracker low pressure off-gas comprises 85% pure hydrogen. By utilizing this in the GOHDS (gas oil hydrodesulphurizer) directly as make-up hydrogen supplement instead of sending it to the hydrogen unit PSA, excess purge gas production is avoided, and precious hydrogen going into the purge gas stream is minimized. It reduced net make-up hydrogen demand and proportionate amount of fuel.

Cost–benefit analysis

Estimated savings due to waste recovery are 3000 MT of LSHS/year, which is equivalent to savings worth Rs 43 million/year. No investment was made. SO_2 emission reduction achieved is 60 MT/year.

Steam traps replacement and repair

One thousand five hundred and seventy traps were replaced, of which 40 HP (high pressure) traps were replaced with TD62, which improved their operational lives. Four hundred and sixty-six traps were repaired.

Cost–benefit analysis

Estimated savings due to replacement and repair are 651 MT LSHS/year, which is equivalent to savings worth Rs 8.58 million/year. Investment made was Rs 1.08 million and payback period is two months. SO_2 emission reduction achieved is 13.02 MT/year.

Solar heaters for canteen/MD bungalow

Solar water heaters of 200-litre capacity replaced electric heaters in the main canteen and MD bungalow.

Cost–benefit analysis

Estimated savings are 17 000 kWh/year, which is equivalent to savings worth Rs 68 000/year. Investment made was Rs 0.18 million and payback period is three years. SO_2 emission reduction achieved is 0.1 MT/year.

Energy saving transformers in SS-22 and SS-23

Energy saving transformers were installed in SS-22 and SS-23. Table 6 gives the net savings after installation (Figure 4).

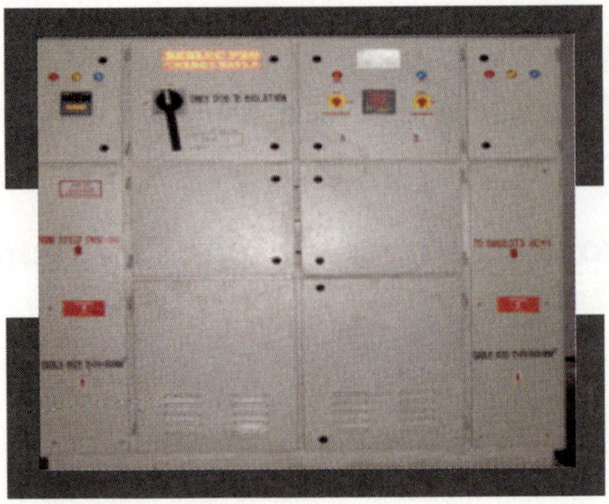

Figure 4
Energy saving transformer

Table 6 Energy savings after transformer installation

Substation	Load before installation (kW)	Load after installation (kW)	Savings (kW)
22	102	82	20
23	127	102	25

Cost–benefit analysis

Estimated savings are 0.19 million kWh/year, which is equivalent to savings of Rs 0.8 million/year. Investment made was Rs 0.64 million and payback period is 10 months. SO_2 emission reduction achieved is 1.9 MT/year.

Process heat integration in crude distillation unit and gas oil hydrodesulphurizer to maximize heat recovery by generating low pressure steam

Brief description

Hot diesel is routed directly from the CDU to the GOHDS. Earlier, feed was heated up to 125°C using stripper bottoms. Now that heat from the stripper bottom is used for heating boiler feed water.

Cost–benefit analysis

The investment made was Rs 3 million. Steam produced is 7 MT/h at full load. Fuel saved in captive power plant is 0.15 MT/h LSHS, which is equivalent to 1200 MT/year. Annual savings achieved are Rs 9.84 million, and payback period is four months.

Benefits

The MRPL has taken several measures to reduce pollution.

- The company maintains lowest specific energy consumption levels through higher capacity utilization, along with the implementation of various energy conservation measures.

- The company is consciously reducing pollution load through optimization of process parameters.
- The company addresses pollution problems through various modifications in process units.

The select projects that have led to reduced pollution and energy conservation are listed below.

- Pollution reduction by quality improvement projects: Isomerization project
- Pollution reduction by loss reduction: volatile organic carbon emission survey
- Pollution reduction with thermal energy optimization: routing amine flash column off gases to sulphur recovery unit incinerator.
- Pollution reduction with electricity reduction: variable speed drives
- Renewable energy utilization: solar lighting and traffic signals
- Waste utilization: biogas plant

CASE STUDY AT A GLANCE

- The MRPL (Mangalore Refinery and Petrochemicals Ltd) is a grass-root refinery, set up as a joint venture company by the Hindustan Petroleum Corporation Ltd and Indian Ryon Industries Ltd.
- The company has achieved a turnover of Rs 373 390 million in 2007/08, registering a net profit of Rs 12 720 million.
- The company has taken various pollution reduction and energy conservation measures.
- All the projects have resulted in substantial savings, along with energy conservation and reduction in the emissions of major air pollutants like oxides of sulphur, oxides of nitrogen, and carbon dioxide.

CASE STUDY 19

Indian Farmers Fertiliser Cooperative Ltd

Figure 1 IFFCO Aonla Unit

About the company

IFFCO (Indian Farmers Fertiliser Cooperative) Ltd is the world's largest fertilizer cooperative based in India, which is registered as a Multistate Cooperative Society. The Society is primarily engaged in the production and distribution of fertilizers. The number of cooperative societies associated with IFFCO have risen from 57 in 1967 to more than 37 700 at present.

Over the years, IFFCO has emerged as a cooperative in true spirit. It is now entirely owned by its cooperatives and has become a cooperative of the farmers, by the farmers, and for the farmers. Initially, IFFCO commissioned an ammonia–urea complex at Kalol and an NPK (nitrogen, phosphorus, potassium)/DAP (di-ammonium phosphate) plant at Kandla, both in Gujarat, in 1975. Another ammonia–urea complex was set up at Phulpur in Uttar Pradesh in 1981. The ammonia–urea unit at Aonla, Uttar Pradesh, was commissioned in 1988. Recently, in September 2005, IFFCO has acquired an NPK/DAP and phosphoric acid fertilizer unit at Paradip in Orissa. The marketing of IFFCO's products is channelled through cooperative societies and institutional agencies in over 28 states and union territories of India.

During 2007/08, IFFCO produced 6.85 MT (million tonnes) of fertilizers, registering an overall capacity utilization of 94% for nitrogenous and 58% for phosphatic fertilizers. It accounted for 20% of country's total nitrogenous fertilizer production and 25% of total phosphatic fertilizer production in the same period. Plant productivity during the year stood at 1354 tonnes/person. IFFCO Aonla Unit produces fertilizer-grade urea, registering an annual production of 1 864 808 MT (Figure 1). The company's turnover in 2007/08 was Rs 121.63 billion.

Environmental profile

IFFCO Aonla Unit is ISO 9001:2000, ISO 14001:2004, and OHSAS 18001:1999 certified.

Environment policy

IFFCO Aonla Unit is committed to conserving the environment and has included environment-friendly practices in all its operations. It has a corporate SHE (Safety, Health, and Environment) Policy at the corporate level and an Environment Policy at the unit level. The policy underscores the following.

- Complying with applicable legislations at the workplace.
- Monitoring and conserving natural resources.
- Minimizing the effluent and waste by recycling and reusing them.
- Generating environmental awareness among employees and other people on the environmental issues.

Environment and Pollution Control Cell

The Unit's commitment towards environment can be gauged from the fact that it has a separate 'Environment and Pollution Control Cell'. The Cell is responsible for ensuring the efficient working of pollution control devices installed in its various plants, compliance with environmental rules and regulations, management of hazardous/biomedical waste, conservation of natural resources, and implementation of environment management programmes adhering to ISI: 14001: 2004 in the organization.

The Cell has an environmental laboratory with modern, state-of-the-art equipment and has experienced and qualified chemists. The laboratory monitors all major pollutants generated by the facility, which helps in ensuring that all these pollutants remain well within the prescribed standards.

Environment Management System

The facility has an EMS (Environment Management System) at the unit level in place. In 2000, the EMS got the ISO: 14001: 1996 accreditation, which was upgraded to ISO 14001: 2004 standards. Among its various activities, the management system focuses on the following.

- Utilization of resources prudently to minimize waste generation.
- Abatement of pollution at the source of generation.
- Conservation of resources.
- Ensuring safe disposal of hazardous waste.
- Continuous monitoring of process parameters for the smooth operation of plants, energy conservation, and environment protection.
- Monitoring of the quality of effluents and air emissions.

Awards

IFFCO has many awards and recognitions in the field of the environment to its credit.

- IFFCO Aonla Unit won the Golden Peacock Environmental Management Award 2008 of the World Environment Foundation, UK.

- The CII (Confederation of Indian Industry) bestowed on the company the National Award for Excellence in Energy Management 2008 and Innovative Project Award.

Air pollution

The major air pollutants released from the Unit include SPM (suspended particulate matter), SO_x (oxides of sulphur), NO_x (oxides of nitrogen), ammonia gas, and fugitive dust and gases.

Air pollution control measures

- Sulphur-free natural gas/regasified liquefied natural gas and naphtha with low sulphur, in the range 25–50 PPM (parts per million), are used as fuels in the primary reformer of ammonia plants and captive power plant.
- Flue gases leaving the furnaces are cooled down to a temperature of 150 °C and vented through a 32-metre-tall chimney in primary reformers and heat recovery units, and through a 120-metre-tall chimney in the steam generation unit in the captive power plant.
- To ensure efficient and complete combustion of fuels, 5% excess air is maintained in the furnaces, which is continuously monitored by an online oxygen analyser.
- A purge gas recovery plant has been set up to recover ammonia and hydrogen from purge gases released from ammonia plants. The recovered ammonia is converted to urea, and hydrogen is reused in ammonia plants for producing ammonia.
- The prill towers in both the urea plants are based on natural drafts. The prill tower is 96-metre high in Aonla-I Unit and 104-metre high in Aonla-II Unit. The prill bucket is so designed and plant-operating parameters are so controlled that the urea dust in the prill tower emissions always remains well below the standard of 50 mg/Nm^3.
- The urea dust generated from the product handling plant is collected, dissolved, and reused in the process concerned with the production of urea.
- Ammonia vapours generated in the ammonia storage tanks are normally diverted to the refrigeration compressor of the ammonia plant. Flare stacks are in place to remove air pollutants generated in ammonia storage tanks.
- Stacks of various heights are provided for controlling and dispersing fugitive emissions.
- Wind socks charged fire hydrant system with reserved fire water capacity of 14 760 m^3, task fire fighting force, water/foam tenders, and smoke detectors with fire alarm system have been provided in various locations.

Table 1 gives the annual average values of the main pollutants and their range observed in ambient air during the last five years.

Water pollution

The waste water released from the facility contains pH, TDS (total dissolved solids), TSS (total suspended solids), BOD (biochemical oxygen demand), COD (chemical oxygen demand), oil and grease, thermal discharge, and heavy metals.

Table 1 Annual average values of the main pollutants released from the Aonla Unit and their range (μg/Nm³)

Parameter	NAAQS	2003/04	2004/05	2005/06	2006/07	2007/08
Ammonia (24 hours)						
Range	400	28–100	15–81	23–160	17–88	28–65
Annual average	100	55.05	51.58	61.55	40.13	42.10
Suspended particulate matter (24 hours)						
Range	500	50–230	80–284	28–397	96–232	141–232
Annual average	360	187.4	164.48	163.92	182.75	192.60
Sulphur dioxide (24 hours)						
Range	120	0.25–11.4	0.1–10.1	0.26–8.71	0.29–2.12	0.1–1.05
Annual average	80	2.78	2.70	2.99	1.84	0.62
Oxides of nitrogen (24 hours)						
Range	120	0.98–49.5	1.0–20	0.21–50.28	0.92–53.10	0.20–1.8
Annual average	80	8.33	4.93	7.27	1.84	1.27

NAAQS – National Ambient Air Quality Standards for the industrial area

Water pollution control measures

- The pH of the effluent discharged from the battery limit of the factory is maintained within the MINAS (Minimal National Standards) limit of 6.5–8.
- Process condensate is stripped in a process condensate stripper to remove the dissolved impurities like methanol, ammonia, and CO_2 from it. The stripped process condensate is reused as boiler feed water after polishing it in the DM (de-mineralization) plant.
- Process condensates from the vacuum section of the urea plant, containing 4%–5% of ammonia and 1%–1.5% urea as impurities, are treated in the urea deep hydrolyser, where urea is decomposed to ammonia and CO_2, which are passed through a distillation column, where ammonia and urea are recovered and reused in the urea plant as feed. The remaining condensate from the distillation column containing <5 PPM ammonia and <1 PPM urea is reused as boiler feed water after polishing in the DM plant.
- The continuous blowdown from boilers from the ammonia and steam generation plants and power generation plant is reused as cooling tower make up.
- The oil is recovered from the oily water/spilled oil and sold to the CPCB (Central Pollution Control Board)-authorized recyclers. The waste water generated is treated in the ETP (effluent treatment plant).
- The waste water containing ammonia is treated in the ETP. This treated water is used for watering the plants in the green belt developed in the premises of the Unit.

- The concentration of heavy metals in the treated effluent is not traceable. Cooling water is treated using non-chromate-based corrosion inhibitors.
- To ensure that the treated effluent has ambient temperature, it is mixed in an equalization pond and is held there for some days.

Table 2 gives the concentration (average and range) of various water pollutants for the past five years.

Table 2 Concentration (average and range) of various water pollutants for the past five years

Parameter	MINAS	2003/04	2004/05	2005/06	2006/07	2007/08
pH	6.5–8					
Average		7.58	7.46	7.52	7.8	7.57
Range		6.9–7.8	7.0–7.7	7.2–7.9	7.3–7.8	7.0–7.9
Total ammoniacal nitrogen as N (mg/l)	50					
Average		13.67	10.67	11.98	11.92	9.32
Range		5.3–29	3.8–17.8	8.1–19.1	1.7–24.9	2.68–16.82
Total Kjeldahl nitrogen as N (mg/l)	100					
Average		23.49	15.45	18.27	24.99	26.74
Range		10.9–38	6.0–25.1	11.2–24.4	6.8–41.9	8.11–41.59
Oil and grease (mg/l)	10					
Average		NT	NT	NT	NT	NT
Range		NT	NT	NT	NT	NT
Total suspended solids (mg/l)	100					
Average		36.33	35.50	18.96	33.33	34.66
Range		22.0–59.0	22.0–52.0	5.5–40.0	18–38	22–50
Vanadium as V (mg/l)	0.2					
Average		NT	NT	NT	NT	NT
Range		NT	NT	NT	NT	NT

MINAS – Minimum National Standards; NT – not traceable

Hazardous waste

The various types of hazardous waste generated by the facility include spent catalyst, spent lube oil, and acid–alkali waste water. Spent catalyst and lube oil are sold to the

CPCB-authorized recyclers. The acid–alkali waste water is neutralized to a pH of 6.5–8.0. This neutralized water is mixed with other treated waste water in equalization ponds and reused for horticultural purposes.

Other waste

- Scrap metals like iron, copper, brass, used industrial filter, and so on are sold to recyclers.
- Torn HDPE (high-density polyethylene) bags and glass bottles are also sold to recyclers.

Case study

Installation of a CDR (carbon dioxide recovery) unit to recover 450 MTPD carbon dioxide from the flue gases of the primary reformer of the Ammonia Plant-I

Background

To bridge the gap between the supply and demand of urea, the Government of India recommended capacity enhancement of the existing units. Under this initiative, CO_2 requirement for producing urea cannot be met only by natural gas, which is used as a feed. The natural gas being supplied to the facility is a lean gas with a low carbon:hydrogen ratio and low calorific value compared to naphtha. So, the plant operation was likely to face CO_2 shortfall during the process of conversion of ammonia to urea.

To address the problem of the shortfall of CO_2, two options were explored.

1. Using naphtha as a supplementary feed, along with natural gas, in the ammonia plant.
2. Setting up a CDR (carbon dioxide recovery) unit for recovering CO_2 from flue gases of the primary reformer (Figure 2).

Figure 2
CDR plant

Technical and financial feasibility studies were conducted to determine the suitability of these options. The operating cost of the first option was found to be very high, even without considering the investment cost. This is because naphtha is very expensive. The study also revealed that the cost of producing CO_2 using naphtha as a feed is 10 times higher than the cost of CO_2 production by setting up a CDR unit. So, the cost-prohibitive first option was rejected, and the facility decided to install a CDR unit. This project also led to the reduction in CO_2, a potent greenhouse gas, emissions to the atmosphere.

Working of the CDR unit

The technology used in the CDR unit is the post-combustion amine scrubbing technology that uses proprietary hindered amine solution (trade name – KS-I) with specially designed equipment.

Various fuels used in the primary reformer of Ammonia Plant-I are natural gas, naphtha, purge gas, and tail gas. Flue gases released from the auxiliary steam super heater, which uses natural gas and naphtha, are mixed with the flue gases released from the primary reformer in the convection section.

Natural gas plus tail gas case was selected as a design basis for the CDR plant, as it has lower CO_2 content and higher moisture content compared to other flue gas cases in the plant.

The flue gases are extracted from the discharge of the ID fan connected to the primary reformer unit and brought to the CDR plant by a flue gas blower through a flue gas water cooler.

In case the flue gas blower or the CDR plant doesn't work, the flue gases are discharged directly to the atmosphere through the existing stacks, so that the operation of the primary reformer unit is not be affected.

The CDR unit of 450 MTPD (metric tonnes per day) CO_2 capacity comprises four sections.

1 Flue gas pretreatment section
2 CO_2 recovery section
3 Solution regeneration section
4 CO_2 compression section

Figure 3 depicts the various sections of the CDR unit.

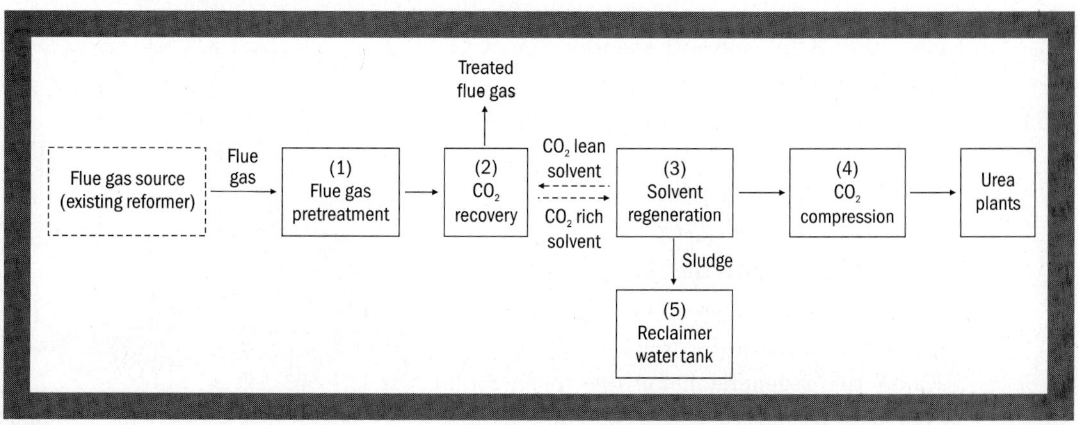

Figure 3 Sections of the CDR unit

Flue gas pretreatment section

The hot flue gases that have a temperature in the range 150–180 °C are cooled to a temperature of 40–45 °C using a flue gas cooling system. The system comprises a flue gas water cooler, which is a packed column–a structured packing to minimize pressure loss. The flue gas is cooled in the upper packing section of the cooler, where it comes in direct contact with water that is supplied from the top of the cooler tower. The flue gases from the primary reformer contain 5 PPM of SO_x. This SO_x is removed in the lower packing section by using a caustic solution. Figure 4 shows the flue gas pretreatment area of the CDR plant.

Figure 4
Flue gas pretreatment area of the CDR plant

CO_2 recovery

The CO_2 absorber (Figure 5) uses the structured packing to reduce the pressure drop of the flue gases inside the CO_2 recovery column. The CO_2 absorber has two sections: CO_2 absorption section (bottom section) and treated flue gas section (top section).

CO_2 absorption section

The cooled flue gases from the flue gas water cooler enter the bottom section of the CO_2 absorber through a flue gas blower. The flue gases move upward through the packing, while the CO_2 lean solution, KS-1 solvent, is introduced from the top of the absorption section. On coming in contact with the flue gases, the KS-1 solvent absorbs the CO_2. The flue gases then move to the treated flue gas washing section in the upper part of the absorber.

Figure 5 CO_2 absorber

Treated flue gas section

The amine content of the flue gases is washed off with water, which also cools these gases. The CO_2-rich solution from the bottom of the absorber goes to the regenerator for the regeneration of the KS-I solvent by a rich solution pumped through a solution heat exchanger.

Solution regeneration/CO_2 stripping

The regenerator is a random packed column, where the rich solution is steam stripped, and CO_2 is recovered from the solution (Figure 6). The overhead vapour is cooled to 40 °C by the regenerator condenser. The partially condensed water is returned from the regenerator reflux drum to the top of the regenerator by the regenerator reflux pump. The CO_2 then enters the CO_2 compressor unit (Figure 7).

Figure 6 CO_2 regenerator

The lean KS-I solvent is cooled to a temperature of 40 °C by the solution heat exchanger, and the lean solution is made cooler prior to being sent to the CO_2 absorber. About 10% of the lean solution is passed through a carbon filter to remove oil and soluble impurities.

CO_2 compression

The CO_2 gas recovered is compressed using a CO_2 compressor and cooled to 40 °C using compressor discharge cooler. This CO_2 is sent to the urea plant.

Solution reclaiming

Residual SO_x and nitrogen dioxide are stronger acids than CO_2 and react with KS-1 solution in the CO_2 absorber to form HSS (heat stable salts). Reclaimer is provided in the CDR unit to remove the HSS formed. When the HSS

Figure 7
CO_2 compressor unit

content in the solvent reaches the limit of 2%, the reclaimer is operated to boil the solution, so that HSS are concentrated to sludge, which is drained to the reclaimed waste tank. The solution returns to the system. Figure 8 gives a schematic diagram of the process.

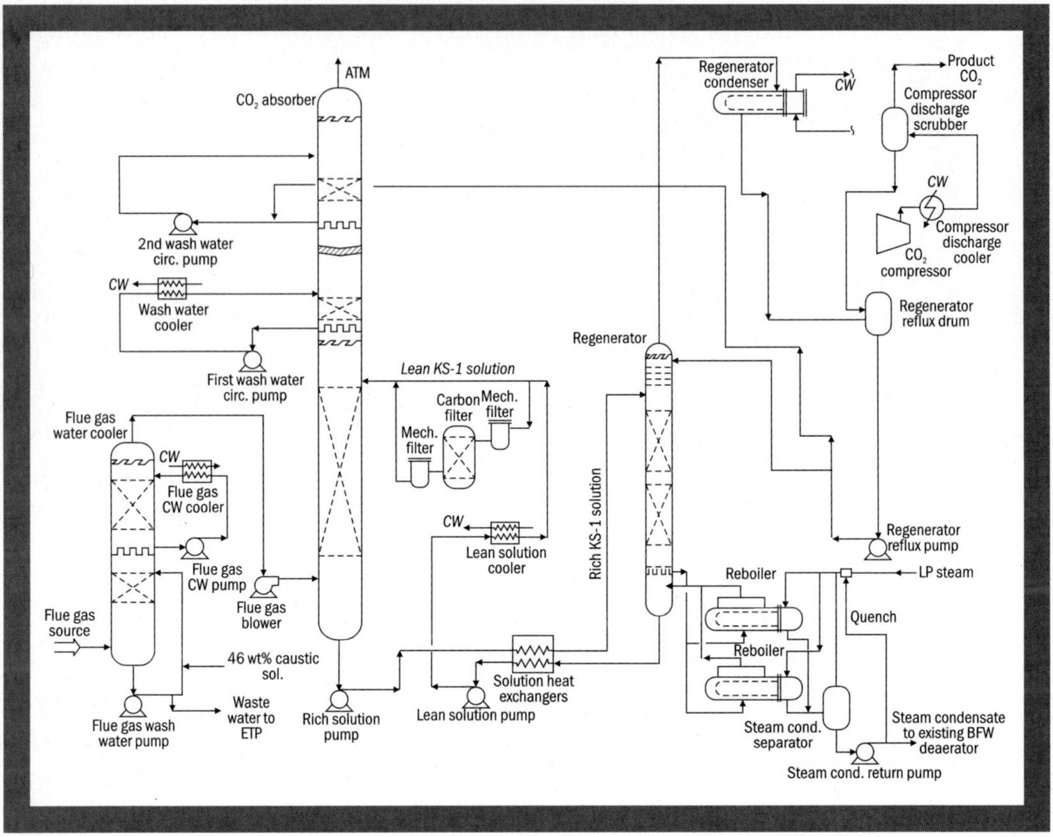

Figure 8 Schematic diagram showing the working of the CDR unit

Benefits

- The CDR plant does not generate any effluent during the process of CO_2 recovery.
- In 2006/07, 18 946 MT CO_2 was recovered from the flue gases of the primary reformer of the Ammonia Plant-1, while 73 991 MT CO_2 was recovered in 2007/08.
- This CO_2 was used for the production of urea, a useful fertilizer.
- The plant has, thus, reduced the emission of CO_2, a potent greenhouse gas.
- During the period 2006–08, 92 937 MT of CO_2 was recovered; 123 915 MT of urea was produced from the recovered CO_2.

CASE STUDY AT A GLANCE

- Indian Farmers Fertiliser Cooperative Ltd, also known as IFFCO, is the world's largest fertilizer cooperative based in India, engaged in the production and distribution of fertilizers.
- The company's turnover in 2007/08 was Rs 121 630 million.
- IFFCO Aonla Unit, a CO_2 (carbon dioxide) recovery, or CDR, unit was installed to recover 450 MTPD CO_2 from the primary reformer flue gases of the Ammonia Plant-I.
- Installing a CDR unit resulted in the reduction of CO_2, a potent greenhouse gas, emissions to the tune of 450 MTPD.
- The recovered CO_2 was used to produce urea–during the period of 2006–08, 92 937 MT of CO_2 was recovered; 123 915 MT of urea was produced from the recovered CO_2.

CASE STUDY 20

United Phosphorus Ltd

Factory at Vapi

About the company

The UPL (United Phosphorus Ltd) was established in 1969 as a small-scale unit. It now has a business office in Mumbai and seven factories in Gujarat, apart from various industrial units in different parts of the world (Figure 1). It produces phosphorus, caustic/chlorine, organo-phosphorus intermediates, organo-phosphorus pesticides, synthetic pyrethroids, sulphonyl urea herbicides, and other pesticides and formulations. Its annual turnover in 2007/08 was Rs 37.61 billion.

Environmental profile

The UPL corporate policy is to implement an environment improvement programme at each and every factory site and business site. All the industrial units are ISO 9001, ISO 14001, and OHSAS 18001 certified. The company has a research unit called Chemo Electronic Laboratory, which does basic research on pollution control instruments and manufactures many such instruments and devices.

Awards/accolades

- National Research and Development Corporation Award 2001 for the invention: 'Chemical detectors for toxic gases, detector strips, short term detector tubes and dosimeter tubes'.
- National Award 2007 from the Department of Science and Industrial Research for outstanding research and development on pollution control and environmental protection.

Air pollution control measures

- To prevent the emissions of SO_2 (sulphur dioxide), the process now uses PCl_5 (phosphorus pentachloride) instead of $SOCl_2$ (thionyl chloride). This helps to recover $POCl_3$ (phosphorus chloride oxide) as a valuable by-product.
- To address the problem of toxic and inflammable phosphine gas produced in the aluminium phosphide plant, fresh air environmental system with scrubber is installed.
- Boilers and thermic fluid heaters are operated on the industrial gas instead of furnace oil, which reduces the emissions of carbon, SPM (suspended particulate matter), and NO_x (oxides of nitrogen).
- All the processes are NO_x emission free.
- To mitigate SPM emission, trees have been planted in the premises and closed and continuous plants ensure that SPM is not released outside.
- The operations in the chemical plants are noiseless.
- Though there is slight chemical odour, it is not noxious.
- For odorous compounds like ethyl mercaptan, there is a closed handling system fume incinerator.

Water pollution control measures

- The effluent released from the facility is directed to the CETP (central effluent treatment plant) (Figures 2 and 3).
- pH of the effluent is neutral.
- There are no suspended solids and heavy metals.
- Oil and grease are removed from the effluent before it is released out.
- The treated water is recycled in cooling towers.

Figure 2 Effluent treatment plant at Vapi

Solid waste and hazardous waste

Solid waste generated is treated and neutralized before being sent to landfill sites at Ankleshwar (Figure 3). Most of the hazardous waste is incinerated. Some by-products are reprocessed and recycled scientifically.

Figure 3 Effluent treatment plant at Ankleshwar managed and run by United Phosphorus Ltd.

Case study

Process improvement to increase the yield of cypermethrin while reducing pollution load, by-product residues, and cost

In all its product lines, the UPL follows the latest technology to reduce environmental pollution and produce quality products. The company has MAXPRO, where 25 chemical engineers work continuously towards process improvement. One of the major products of the company is cypermethrin, which is a synthetic pyrethroid pesticide, the highest selling pesticide in the world (Figure 4).

Process improvement

The basic raw materials used to produce cypermethrin are MPBAD (meta phenoxy benzaldehyde) and DVACL (dichlorovinyl acid chloride). Both the raw materials are manufactured in the plant located at Vapi. By continuously improvising, the yield has been improved by 8%. Aluminium chloride, which is generated as a by-product, is converted to aluminium hydroxide (alumina gel), which is a pharmaceutical product. The other by-product potassium bromide is converted to bromide, which is reused in the plant. So, the company constantly endeavours to minimize or reuse by-products.

After extensive in-plant research and development, the yield of DVACL has been increased by more than 10%, which has resulted in less by-product residues and less load on waste disposal system. Also, in the new developed process, the polluting $SOCl_2$ has been replaced

Figure 4 Cypermethrin plant in Vapi

with PCl_5. Apart from improving yield and reducing pollution, it gives a valuable by-product, $POCl_3$.

Cypermethrin was normally produced in a batch process. After conducting some research and development studies, the process was changed to a continuous process, which has again reduced by-products and pollution, and enhanced the quality of the product. Also, continuous distillation of the final product has reduced the generation of by-products.

Because of process improvement, the company has upscaled its capacity from 100 tonnes to 500 tonnes per month. The company has also managed to reduce the size of its equipment, which has led to huge savings and reduction in the cost of the final product. For example, instead of using 15 kl reactors in batch process, the company is now using 3 kl reactors.

When cypermethrin was imported in India, it used to be 92% pure. The company now makes 95%–98% pure product. Also, India used to import cypermethrin at $100 per kg from the US and Europe. But now, the UPL is exporting it to the same countries at less than $10 per kg.

Benefits

- *Environment friendly* The process of manufacturing cypermethrin was changed to a continuous process, which has reduced by-product generation and pollution.
- *Increase in capacity* Because of process improvement, the capacity of the company has upscaled from 100 tonnes to 500 tonnes per month.
- *Cost cutting* The size of the equipment has also reduced, which has led to huge savings, cutting down the cost of the final product.
- *Purity of product* Because of process improvement, the product is now 95%–98% pure.

CASE STUDY AT A GLANCE

- The UPL (United Phosphorous Ltd) was established in 1969 as a small-scale unit.
- It produces phosphorus, caustic/chlorine, organo-phosphorus intermediates, organo-phosphorus pesticide, synthetic pyrethroids, sulfonyl urea herbicides, and other pesticides and formulations.
- Its annual turnover in 2007/08 was Rs 37.61 billion.
- Process improvement has resulted in the increase in the yield of cypermethrin, while reducing pollution load, by-product residues, and cost.
- Because of the process improvement, the company has upscaled its capacity from 100 tonnes to 500 tonnes per month.
- The company is exporting cypermethrin to the US, Europe, Africa, Latin America, and Far East countries.

CASE STUDY 21 Orchid Chemicals and Pharmaceuticals Ltd

About the company

Orchid Chemicals and Pharmaceuticals Ltd, or Orchid, was established in 1992 as a 100% export-oriented unit. Commencing operations in 1994, Orchid has achieved consistent growth, both quantitatively and qualitatively, to emerge among the top 15 companies in the Indian pharmaceutical industry in a short span of 14 years of its operations.

Orchid is today a globally recognized, integrated pharmaceutical company with core competencies in the development and manufacture of APIs (active pharmaceutical ingredients) and finished dosage forms as well as in drug discovery. From its very inception, Orchid has been investing aggressively for establishing modern research and manufacturing facilities aimed at global markets, thus emerging as a world-class pharmaceutical company covering the entire value chain from 'discovery to delivery'.

Orchid has two manufacturing sites for APIs (at Alathur, near Chennai, and at Aurangabad, near Mumbai) and three manufacturing sites for dosage forms (at Irungattukottai and Alathur in Chennai), besides two R&D centres (at Sholinganallur and Irungattukottai, Chennai).

Orchid's scientific and technical strengths have made it a partner of choice for several multinational corporations. Orchid has long-term exclusive marketing alliances with reputed global companies such as Apotex, Actavis, Dava, and Hospira for the distribution of Orchid's products in the advanced markets of the US and Europe.

Orchid has an established end-to-end connected infrastructure for drug discovery and development, channelled through its two subsidiaries, Orchid Research Laboratories in Chennai and Bexel Pharmaceuticals in the US.

Environmental profile

Orchid constantly works towards making its processes more energy efficient and reducing their adverse environmental impacts. It has a combined Safety, Health, and Environmental Policy, which underscores the need for continually improving environmental management systems by focusing on loss control, prevention of pollution, elimination of hazards, and meeting legal and other requirements. Orchid's API facilities are ISO certified for their quality, environmental management, and operational health and safety systems. It is the youngest Pharma Company in India to receive the coveted ISO 9001:2000 (earlier ISO 9002) certification from Intertek. It is also the first pharmaceutical company in India to have attained the coveted ISO 14001 quality certification from the Dutch Council for Accreditation for its quality environmental management systems.

Air pollution and its control measures

The main air pollutants emitted from the facility include SPM (suspended particulate matter), RSPM (respirable suspended particulate matter), SO_2 (sulphur dioxide), NO_x (oxides of nitrogen), and fugitive dust and gases. All these parameters are monitored on regular intervals at the MoEF (Ministry of Environment and Forests)-approved laboratory. All the process emissions are vented through process scrubbers of adequate size and capacities. All the DGs (diesel generators) and boilers are provided with stacks of appropriate height. Some of the measures taken to mitigate air pollution are as follows.

- *Dust collection system* This system comprises ducts in various rooms and hoods on various equipment, which are connected to a series of filters. The filtered air is recycled back to rooms.
- *Air-conditioned and dehumidified supply and return system* This system comprises an elaborate system of dehumidification, conditioning, and temperature control of the air, with a series of filters.
- *Stacks* The boiler stack is 30.5 m high and the DG stack is 17 m tall. These stacks, along with the scrubber stack, ensure efficient dispersal of flue gases.

Water pollution and its control measures

The waste water discharged from the facility contains BOD (biochemical oxygen demand), COD (chemical oxygen demand), pH, TDS (total dissolved solids), TSS (total suspended solids), and oil and grease. These parameters are monitored in-house in a state-of-the-art laboratory, and care is taken to ensure that all the parameters are within the specified standards. The company segregates streams into low pollution stream and high pollution stream. The low pollution stream is given exhaustive treatment in the effluent treatment plant. The high pollution stream with high TDS and COD is treated in a primary system comprising a flocculator and a clarifier. The company's waste water treatment works on the concept of zero discharge.

Hazardous waste

Since Orchid is involved in bulk drug manufacturing, a lot of organic chemicals are used in the process. As a result, different types of hazardous waste are generated, which include used/spent oil, waste/residue containing oil, spent solvents, distillation residues, spent carbon, spent mother liquor, spent organic solvents, discarded containers/barrels/liners, chemical sludge from waste water treatment plants, and so on. Apart from these, e-waste like printed circuit board and monitors is also generated. Some of these waste items like used/spent oil and waste/residue containing oil are reused and recycled as per the hazardous waste rules. Others like distillation residues, spent carbon, and chemical sludge from the waste water treatment plants are either incinerated or landfilled.

Case study

Catalytic dehalogenation process: Orchid's initiative towards adopting an eco-friendly process

Dehalogenation process is used by Orchid Chemicals to convert 6-APA (6-aminopenicillanic acid) compound to Debromo compound (Figure 1). In the dehalogenation process used earlier by the facility, magnesium metal was used in the presence of hydrochloric acid. This reaction was highly exothermic and required a large quantity of magnesium metal. To control the high temperature conditions, a chilling utility (–40 °C) was used. It was a time consuming process, consuming a lot of energy as well. Moreover, a large quantity of effluent, high on TDS, was also generated, which required additional treatment. Figure 2 gives the outline of the dehalogenation process carried out using magnesium metal.

Figure 1
Hydrogenation reactor

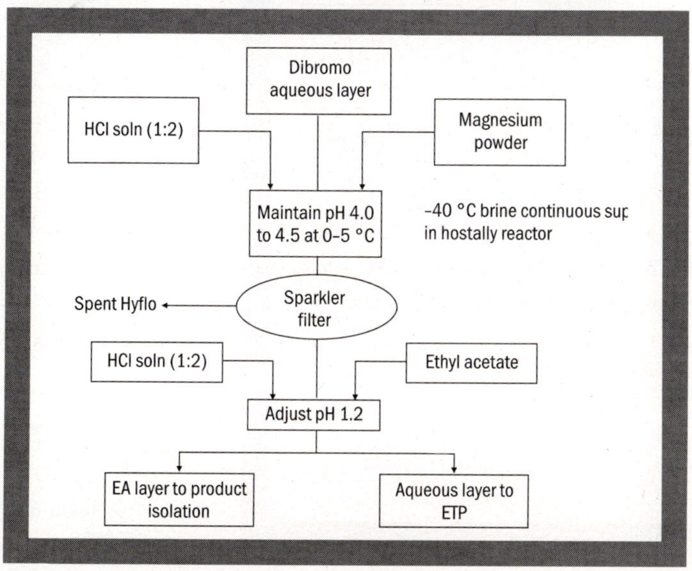

Figure 2
Dehalogenation process using magnesium metal

Because of these drawbacks in the conventional dehalogenation process, the company switched to the catalytic dehalogenation process. In this process, dehalogenation is carried out by catalytic reduction by using palladium/platinum on carbon and hydrogen gases, instead of magnesium metal (Figure 3). Figure 4 briefly describes the catalytic dehalogenation process.

Catalytic dehalogenation does not use magnesium and, therefore, reduces the salt generation (hazardous waste). Moreover, the consumption of hydrochloric acid also reduces, which subsequently reduces the amount of effluent generated. Table 1 gives a comparative account of the two processes.

Figure 3
Catalyst charging through closed system slide

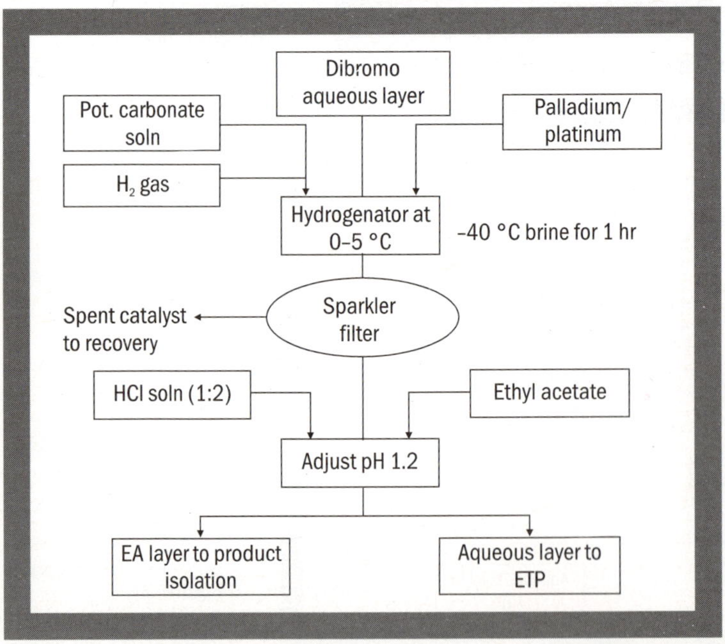

Figure 4
Catalytic dehalogenation process

Table 1 Comparison between the dehalogenation process using magnesium metal and catalytic dehalogenation process

Process	Input quantity per batch	Output quantity per batch	Effluent generated per batch	Salt generated per batch	Batch cycle time per batch	Assay of product	Utility (–40 °C) run time per batch
Dehalogenation using magnesium metal	100 kg	68 kg	1600 litres	622 kg	65–70 hours	92%	24 hours
Catalytic dehalogenation	100 kg	73 kg	1200 litres	369 kg	40–42 hours	96%	1 hour

Cost–benefit analysis

Savings due to the reduction in high pollution effluent stream: Rs 249 600
Savings due to the reduction in hazardous waste generation: Rs 24 298
Savings due to 23 hours reduction in electricity load during the running of utility: Rs 1 346 990
Savings due to the reduction of batch run time: Rs 83 664
Savings in raw material: Rs 384 000
Total savings: Rs 2 088 552, that is, approximately, Rs 2.1 million [annual figures]

Benefits

Switching to the catalytic dehalogenation process has resulted in the following benefits.

- There has been a 40% reduction in the generation of hazardous waste.
- There has been a 25% reduction in the effluent generated.
- Furnace oil consumption has reduced, which leads to 15.5 kg/annum reduction in sulphur emissions.
- Electricity consumption has reduced by 95%.
- Productivity has improved by 55%.
- There is a complete elimination of unrecoverable metal.
- The total savings per annum are to the tune of Rs 2.1 million.

CASE STUDY AT A GLANCE

- Orchid Chemicals and Pharmaceuticals Ltd, or Orchid, was established in 1992 as a 100% export-oriented unit.
- Orchid is today a globally recognized, integrated pharmaceutical company with core competencies in the development and manufacture of APIs (active pharmaceutical ingredients) and finished dosage forms as well as in drug discovery.
- In an attempt towards greening its processes, Orchid has adopted an eco-friendly process, the catalytic dehalogenation process.
- The catalytic dehalogenation process has resulted in a 40% reduction in the generation of hazardous waste.
- There has been a 25% reduction in the effluent generated.
- Furnace oil consumption has reduced, which leads to 15.5 kg/annum reduction in sulphur emissions.
- Electricity consumption has reduced by 95%.
- Productivity has improved by 55%.
- There is a complete elimination of unrecoverable metal.
- The total savings per annum are to the tune of Rs 2.1 million.

CASE STUDY 22

Ispat Industries Ltd

About the company

Ispat Industries Ltd, or IIL, is one of the leading integrated steel-makers and the largest private sector producer of hot rolled coils in India. Set up as Nippon Denro Ispat Ltd in 1984 by its founding chairman Mr M L Mittal, the IIL has steadily grown into a Rs 94-billion company, assuming its position as the flagship company of the reputed Ispat Group. A corporate powerhouse with operations in iron, steel, mining, energy, and infrastructure, the Ispat Group today figures among the top 20 business houses in the country.

Headquartered at Mumbai, the IIL is the leader in the national speciality steel market. The company's core competency is the production of high quality steel, for which it employs cutting-edge technologies and stringent quality standards. It produces world-class sponge iron, galvanized sheets, and cold rolled coils, in addition to hot rolled coils, at its two state-of-the art integrated steel plants located at Dolvi and Kalmeshwar in the state of Maharashtra.

The sprawling 1200-acre Dolvi complex houses the 3-MTPA (million tonne per annum) hot rolled coils plant, which combines the latest technologies – the Conarc process for steel making and the compact strip process – introduced for the first time in Asia.

The complex also has a 1.6-MTPA sponge iron (DRI) plant. Moreover, the Dolvi complex is home to a 2 MT blast furnace. A new 2.24-MTPA sinter plant, a 1260 tonnes per day oxygen furnace, and a new electric arc furnace have also been commissioned at IIL, Dolvi.

The Kalmeshwar complex houses Ispat's 0.4-MT cold rolling complex, which also includes the galvanized plain/galvanized corrugated lines and India's first colour coating mill.

Environmental profile

The IIL, as a responsible corporate citizen, is committed to conserving and preserving the environment by adopting best practices in all its operations. To ensure that all its operations are environment friendly, the company has set up a separate environment department. The department has a well-established laboratory with state-of-the-art pollution testing equipment for monitoring water, air, and noise quality. The department also ensures that the company complies with all the environmental laws and regulations stipulated by the Maharashtra Pollution Control Board, Ministry of Environment and Forests, and International Finance Corporation.

The department is actively involved in the implementation of the provisions of ISO 14001 certification for the sponge iron plant.

The company has a proactive environment policy at both corporate and facility levels. The policy outlines the following guidelines for the company.

- Commitment to conform to national and international standards and striving towards continually improving the environmental performance of the company.
- Optimizing resource utilization, energy conservation, and waste management through pollution prevention initiatives.

- Implementation of effective Environment Management System by setting and periodically reviewing relevant objectives and targets.
- Addressing concerns related to the environment, safety, and health of the employees.

Air pollution

The major pollutants generated from the facility are SPM (suspended particulate matter), RSPM (respirable suspended particulate matter), SO_2 (sulphur dioxide), NO_x (oxides of nitrogen), and CO_2 (carbon dioxide).

Air pollution control measures

Efforts are made to reduce air pollutants at the source. This is achieved by installing in-built pollution control systems at strategic points. Some of these systems are listed below.

- Dust suppression system for the conveyor
- Dedusting system at the stock house and coal handling unit
- Dust catcher
- Cast house fume extraction system
- Gas cleaning plant
- Tall stacks to ensure the dispersal of pollutant gases

Water pollution and its control measures

The major pollutants generated include COD (chemical oxygen demand), pH, oil and grease, and TDS (total dissolved solids). A waste water treatment plant is in place for treating waste water generated from the blast furnace. The treated water is recirculated back into the process.

Hazardous waste

The main hazardous waste generated is used oil/spent oil, which is sold to the CPCB (Central Pollution Control Board)-authorized vendors for recycling.

Other waste

Slag, sludge generated from the gas cleaning plant, and raw materials like coke fines and oxide fines are the other types of waste generated in the facility. These are disposed of in the following manner.

- Slag generated is used in cement plants for making cement.
- The sludge generated from the gas cleaning plant is used for making sinter.
- Raw material fines like coke fines and oxide fines are used for making sinter.

Case study

Conservation of energy by the recycling of blast furnace waste gas as a source of heat supply in the sinter plant, coal injection plant, and Indorama Cement Plant

Generation of blast furnace gas

The blast furnace at the IIL is a self-supporting structure, internally lined with refractories (Figure 1). It produces molten metal. Raw materials like sinter, iron ore, coke, and coal are charged from the top of the furnace, and hot blast is charged from the bottom through tyures. The iron ore gets reduced in the presence of coke, and the liquid metal produced is removed from the bottom. The gases produced during the process comprise CO (carbon monoxide), nitrogen, CO_2 (carbon dioxide), and so on (collectively referred to as the blast furnace gas), which are cleaned in the gas cleaning plant. Approximately, 350 000 Nm^3/h of gas is produced in the furnace. This gas has a calorific value of 700 $kcal/Nm^3$. Of the total gas generated, 140 000 Nm^3/h of the gas was being recycled for stove heating, and the rest was being flared in flare stacks. To address this problem of so much gas going waste, the company decided to innovate and use the surplus gas being flared in its various operations.

Figure 1
Blast furnace

Various uses of the blast furnace gas

Use of the blast furnace waste gas in the coal injection plant

The blast furnace gas with a high calorific value is used in the coal injection plant for heating coal to remove its moisture. The furnace gas is ignited with propane in a hot air generator. This ignited gas that has a temperature of 200 °C is transferred to the coal grinding mill, where it is used as a source of heat. The plant uses 2500 Nm^3 of gas per hour. Earlier the plant was designed to use only propane. Now the propane is used for ignition and for stabilizing the flame (Figure 2).

Figure 2
Use of the blast furnace gas in the coal injection plant

Use of blast furnace waste gas in the sinter plant

Sinter is a clinker-like aggregate, normally produced from fine iron ore mixed with coke breeze, limestone dolomite fines, and other metallurgical return waste. These aggregates are ignited in an ignition furnace at a temperature of 1150 °C. The fuel in the ignition furnace is the blast furnace gas that is ignited with propane in the ratio 95:5. About 2500 Nm^3/h of the blast furnace gas is used in the sinter plant (Figure 3). Figure 4 gives the blast furnace gas line in the sinter plant.

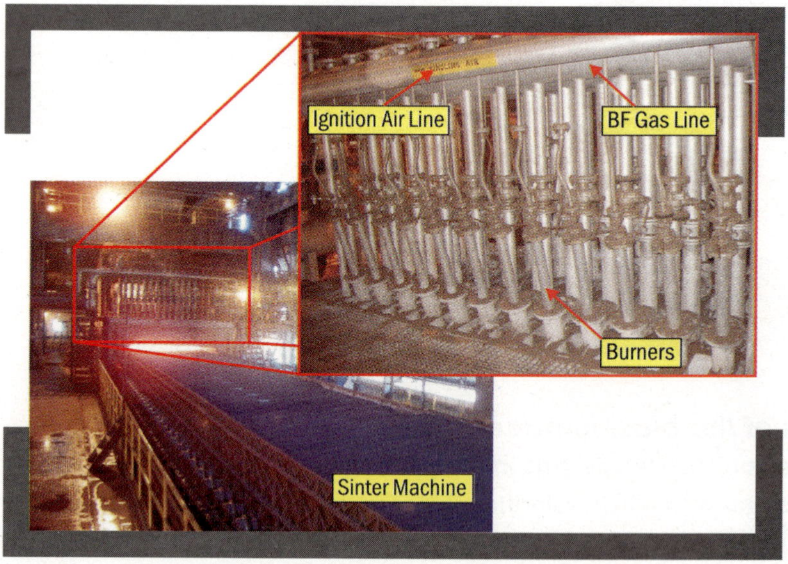

Figure 3
Use of the blast furnace gas in the sinter plant

Use of the blast furnace waste gas at the Indorama Cement Plant

The Indorama Cement Plant located in the plant premises buys 18 000 Nm^3/h of the blast gas from the facility, which is used as fuel. Figure 5 shows the blast furnace gas line at the Indorama Cement Plant.

Figure 4
Blast furnace gas line in the sinter plant

So, altogether, additional 23 000 Nm³/h of the gas is being utilized, and the rest (about 187 000 Nm³/h) is being flared.

Cost–benefit analysis

Saving of propane: 50 Nm³/h
Cost of propane per kg: Rs 31
Annual saving: Rs 6 428 160

Benefits

- *Reduced pollution* The blast furnace gas comprising CO, CO_2, and nitrogen, which was being flared, is now used as a fuel in various operations. This has resulted in a reduction in air pollution.
- *Conservation of fuel* Approximately, 50 Nm³/h of propane is being saved, which has resulted in an annual saving of Rs 6 428 160.

Figure 5
Blast furnace gas line at the Indorama Cement Plant

CASE STUDY AT A GLANCE

- Ispat Industries Ltd, or IIL, is one of the leading integrated steel-makers and the largest private sector producer of hot rolled coils in India.
- The IIL has steadily grown into a Rs 94-billion company, assuming its position as a flagship company of the reputed Ispat Group.
- The blast furnace generates gases like carbon monoxide, carbon dioxide, and nitrogen, collectively referred to as the blast furnace gas.
- Part of this gas was being recycled and the rest was being flared.
- The company developed innovative ways to use this waste gas as a source of heat in the sinter plant, coal injection plant, and Indorama Cement Plant.
- This project has led to the reduction in air pollution and conservation of fuel.
- The saving in the fuel propane has resulted in an annual saving of Rs 6 428 160.

CASE STUDY 23

Chennai Petroleum Corporation Ltd

About the company

Chennai Petroleum Corporation Ltd, or CPCL, formerly known as Madras Refineries Ltd, or MRL, was formed as a joint venture in 1965 between the Government of India, AMOCO, and the National Iranian Oil Company, with a shareholding in the ratio 74%: 13%: 13%, respectively. It is located in Manali in Tamil Nadu. From the grass-roots stage, the CPCL Refinery was set up with an installed capacity of 2.5 MTPA (million tonnes per annum) in a record time of 27 months at a cost of Rs 430 million.

The CPCL has two refineries with a combined refining capacity of 10.5 MTPA. The Manali Refinery has a capacity of 9.5 MTPA and is one of the most complex refineries in India, with fuel, lube, wax, and petrochemical feedstock production facilities (Figure 1). The CPCL's second refinery is located at Cauvery Basin at Nagapattinam.

The main products of the company are LPG (liquefied petroleum gas), motor spirit, superior kerosene, aviation turbine fuel, high-speed diesel, naphtha, bitumen, lube base stocks, paraffin wax, fuel oil, hexane, and petrochemical feedstocks. The crude throughput for 2007/08 was 10.266 MT. The company recorded the highest-ever turnover of Rs 328.89 billion in 2007/08.

Figure 1
CPCL Manali Refinery

Environmental profile

The CPCL has an environment department that operates at both corporate and facility levels. The department has the onus of ensuring that all the operations of the company conform to environmental regulations and standards. There is also an Environment Management System in place.

The CPCL Manali Refinery was awarded ISO 14001 certification by M/s BVQi. The company was also awarded ISO 9001-2000 certification for Quality Management System and OHSAS 18001 certification for Occupational Health and Safety Management System by M/s Det Norske Veritas of the Netherlands.

Air pollution

The major air pollutants emitted from the facility include SO_2 (sulphur dioxide), NO_x (oxides of nitrogen), SPM (suspended particulate matter), RSPM (respirable suspended particulate matter), and CO (carbon monoxide).

Air pollution control measures

The company has implemented various pollution control measures to check the pollutants going into the atmosphere. Some of them are listed below.

- A new hydro-cracker unit has been installed in the expansion refinery to reduce sulphur emissions from the existing and proposed refineries.
- The refinery uses low-sulphur fuel.
- Low-NO_x burners are used in boilers and process furnace heaters to reduce NOx emissions.
- DeNO_x technology has been adopted in major furnaces and heaters.
- Diesel hydro desulphurization project has been implemented to reduce the sulphur levels in the diesel product from 1% to 0.25%, and further down to 0.05% to meet Bharat-II fuel norms.
- Unleaded petrol products are produced to contain the exhaust pollution from automobiles.
- Air preheaters and economizers have been installed to reduce flue gas emissions.
- Floating roof tanks with secondary seals have been provided for crude and light end products to reduce hydrocarbon and fugitive emissions.
- The flare gas recovery unit recovers hydrocarbons routed to the flare system.
- A new sulphur recovery plant with Max Clauss process (sulphur recovery >99%) has been installed to recover elemental sulphur from acid gases.
- Stack heights have been increased to ensure effective dispersion of air pollutants.
- Air quality monitors have been installed in various locations in the Refinery complex for continuous monitoring of ambient air quality, with centralized data acquisition system.

Water pollution and its control measures

By using advanced water treatment technologies, the company has managed to achieve zero discharge status.

- Membranes in the RO (reverse osmosis) plant were replaced with new generation membranes, enhancing throughput and permeate production.
- Improvement in the RO permeate quality has reduced the generation of effluents.
- A new zero discharge plant of 200 kl/h capacity has been commissioned to recycle the

treated effluents. The effluent discharge from the Refinery is nil due to the installation of zero discharge projects.
- By recycling 100% of treated effluents in various peripheral and process applications, huge quantity of raw water is saved.
- The CPCL has installed a sewage reclamation facility, which converts the city sewage into high-quality (boiler feed) water.

Hazardous waste

Hazardous waste generated from the facility includes spent catalysts and oily sludge. The sludge is subjected to mechanical centrifugation, chemical treatment, and bioremediation. This treated sludge is sold to major cement industries, which use this sludge in their furnaces. Spent catalysts are sold to the CPCB (Central Pollution Control Board)-authorized recycling/ reprocessing companies. Also, hydro finishing process used for wax polishing reduces hazardous waste generation.

Case study

Use of wind energy for the operation of the sea water desalination plant

With the world waking up to the adverse impacts of fossil fuels, manifested in the form of global warming, more and more emphasis is being placed on the use of renewable sources of energy. Wind energy as a renewable energy is gaining prominence worldwide. The Indian wind energy sector has an installed capacity of 7082 MW (as on 31 March 2007). Tamil Nadu is a leading state in terms of exploiting wind energy. This is evident from the fact that of the total wind energy generation in India, Tamil Nadu accounts for more than 55%.

To use and promote clean energy, the CPCL has established a large windmill power project of 17.6 MW capacity at Palghat (Pushpathur) Pass, near Coimbatore (Figure 2). Twenty-two wind energy generators, each with 0.8 MW capacity, are operating in this windmill farm. This wind energy conversion project is connected to the TNEB (Tamil Nadu Electricity Board) grid and is providing electricity for the operation of the 5.8-MGD (million gallons per day, or 24 million litres per day) capacity sea water desalination plant. The windmill farm is expected to generate 36 million kWh of electricity per annum.

The sea water desalination project is located near Ennore in north Chennai, which was constructed at a cost of Rs 2300 million (Figure 3). This project fulfills the water demand of the Refinery. In addition, 5.8 MGD of raw water from Metro water, an agency responsible for managing sewage in Manali area, can be diverted for city usage. The desalination plant requires 54 million kWh of electricity annually for its operation. The CPCL meets part of this power demand from wind farm during four months of its operation and the rest from the TNEB, especially during the period when wind farm is not operational. The grid infrastructure is used to wheel electricity from the wind farm to the desalination plant.

Thus, this project combines the issue of water scarcity with sustainable usage of wind energy.

Figure 2
CPCL's windmill farm at Pushpathur

Figure 3
Intake well and pump house for the sea water desalination plant

Benefits

- The annual CDM (Clean Development Mechanism) of the project will be about 3200 CERs (certified emission reductions).
- Use of renewable source for electricity generation ensures that there is no emission of greenhouse gases that are released when fossil fuels are burnt.
- The desalination project will address the water scarcity problem in the area and provide uninterrupted supply of water to the facility.

CASE STUDY AT A GLANCE

- Chennai Petroleum Corporation Ltd, or CPCL, formerly known as Madras Refineries Ltd, or MRL, was formed as a joint venture in 1965.
- The CPCL has two refineries with a combined refining capacity of 10.5 MTPA.
- The company recorded the highest-ever turnover of Rs 328.89 billion in 2007/08.
- The company has developed an innovative project to use wind energy for the operation of the sea water desalination plant.
- A 17.6-MW capacity wind farm has been established by the company, which generates 36 million kWh of electricity per annum.
- The annual CDM (Clean Development Mechanism) of the project will be about 3200 CERs (certified emission reductions).
- Use of renewable source for electricity generation ensures that there is no emission of greenhouse gases that are released when fossil fuels are burnt.
- The desalination project will address the water scarcity problem in the area and provide uninterrupted supply of water to the facility.

CASE STUDY 24 Tamil Nadu Newsprint and Papers Ltd

Figure 1
The Tamil Nadu Newsprint and Papers Ltd at Kagithapuram, District Karur, Tamil Nadu

About the company

The Tamil Nadu Newsprint and Papers Ltd, or TNPL, is situated at Kagithapuram, District Karur, Tamil Nadu (Figure 1). It was established by the Government of Tamil Nadu in the early 1980s to produce newsprint, and printing and writing paper using bagasse, a sugar cane residue, as primary raw material. The TNPL produces about 230 000 MT (million tonnes) of printing and writing paper newsprint per annum, and about one-fifth of its production is exported to more than 30 countries.

Environmental profile

Environmental protection and management are integral parts of the TNPL's activities. These are reflected in the company's vision, which states that the company aspires to be the market leader in the manufacture of world-class eco-friendly paper by adopting innovative technologies for sustainable development. The TNPL has taken several measures for protecting the environment in line with its commitment for sustainable development. In its choice of raw materials, sources of energy, and production processes, as well as in product development, the objective is to minimize the pollution load on the environment.

The company is committed to producing eco-friendly quality papers by adopting innovative technologies and providing a cleaner environment by sustained research and development efforts and continual process improvements. The company complies with all relevant environmental legislations and regulations and conserves and optimizes the use of resources.

By using bagasse as primary raw material, TNPL preserves over 40 000 acres of forest land every year. The company's Environmental Management System is ISO: 14001: 1996 certified, which has been upgraded to ISO 14001: 2004.

Air pollution

The emissions from the facility contain SPM (suspended particulate matter), RSPM (respirable suspended particulate matter), SO_2 (sulphur dioxide), and NO_x (oxides of nitrogen). Fugitive emissions also form a substantial part of the emissions.

Air pollution control measures

- All boiler and kiln stacks are provided with tall chimneys with ESPs (electrostatic precipitators).
- Water sprinklers are used to spray water in the coal handling unit and bagasse handling plant to control fugitive dust emissions.
- Dust extraction systems are also used to check dust emission.
- Bag splitting machine, along with pneumatic conveying arrangement, is also used to mitigate pollution from dust.
- A green cover has been developed around the factory premises, which acts as a natural sink.

Water pollution

The main pollutants discharged in the waste water are BOD (biochemical oxygen demand), COD (chemical oxygen demand), pH, TSS (total suspended solids), TDS (total dissolved solids), and oil and grease.

Water pollution control measures

The facility has a waste water treatment plant to treat its water pollutants, which uses primary and secondary treatment processes. The treated effluent conforms to the standards for discharge into the surface water, as prescribed by pollution control boards. About 1600 acres of arid land is irrigated using the treated effluent. The treatment plant is being augmented and upgraded further to enhance its efficiency.

Hazardous waste

Hazardous waste from the facility is generated in the form of waste oil, effluent treatment plant sludge, black liquor dry solids, and used batteries. Waste oil and used batteries are sold to authorized recyclers, while sludge and liquor dry solids are used as fuel in the plant.

Other waste

The facility also generates waste like dust from chipper house, boiler ash, fly ash, effluent sludge, lime sludge, pith, lime grits, and hypo mud. Dust from chipper house and pith are used as fuels in boilers. Boiler ash and fly ash are used in cement manufacturing. Lime grit is used for constructing roads, while effluent sludge is used as raw material in small cardboard industries.

Case study

Installation of a high-rate biomethanation waste water treatment plant for greenhouse gas mitigation and renewable energy generation

Bagasse utilization

At full production, the company uses nearly 1 000 000 MT sugar cane bagasse per annum to produce bagasse pulp in the biogas plant (Figure 2). The bagasse received from the sugar

Figure 2
Biogas plant

mill, having 3%–4% residual sugar, is stored in the open yard for three to nine months by using classical wet bulk storage method. During storage, the bagasse is kept under wet condition by spraying water over bagasse pile to preserve bagasse quality. After storage, the bagasse is reclaimed and washed to remove residual sugar and depithed to remove the pith. It is then fed into continuous digester/refiner to produce pulp. Figure 3 depicts how bagasse is utilized to produce pulp and other by-products.

Effluent treatment in anaerobic lagoons

The Bagasse washing/storage yard and fibre preparation plant generate effluent with high COD. Earlier, this effluent used to be pretreated in anaerobic lagoons, where the COD level was brought down by 85%. It was then taken to the activated sludge treatment unit, along with the waste water streams generated from other units. Though the COD concentration of the effluent reduced in the lagoons, the uncontrolled anaerobic reaction used to release methane, a potent GHG (greenhouse gas).

Use of a high-rate biomethanation plant to treat effluent

To address this problem of air pollution, a high-rate biomethanation plant replaced anaerobic lagoons, which is essentially an anaerobic reactor based on upflow anaerobic sludge blanket

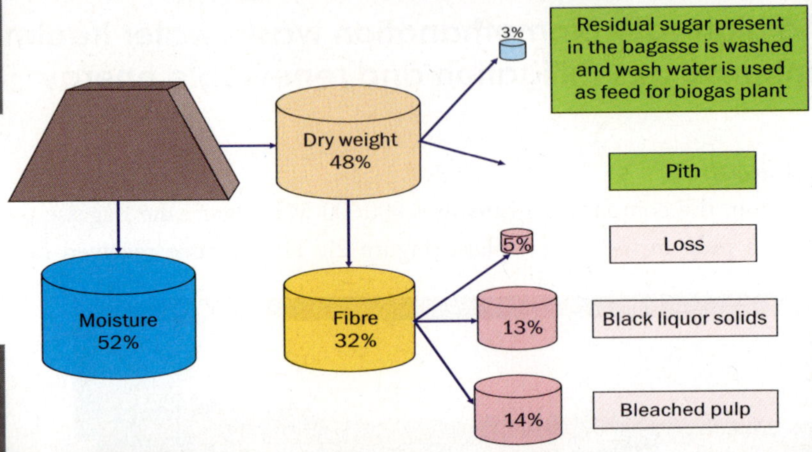

Figure 3 Bagasse utilization in paper-making process

Figure 4 Biomethanation plant

process (Figure 4). The plant is designed to handle 12 000 m^3/day of bagasse wash waste water having 6000 mg/litre of COD, with a hydraulic retention time of 20 hours and maximum loading rate of 5.75 kg COD/m^3/day. The bagasse wash waste water is received from the bagasse clarifier into an equalization tank, from where it is fed to a neutralization tank. Here, the waste water is neutralized using milk of lime so as to raise the pH of the waste water from 4–4.5 to 6.5–7. Following neutralization, the waste water is clarified in the clarifier to remove suspended solids. The settled sludge is taken to the sludge pit and then to the decanter centrifuge to separate solids. The separated solids are used as fuel in the power boiler after solar drying. The clarified waste water is taken to the buffer tank, where nutrients like urea and di-ammonium phosphate are added. From the buffer tank the waste water is fed to two reactors, and the treated effluent from the reactors is subjected to activated sludge process. Biogas is generated in the reactors, which is stored in a gasholder, from where it is pumped to lime kiln using gas blowers. This biogas comprises 70%–75% methane and has 6400 kcal/Nm^3

of heat value. The various processes taking place in a biomethanation plant are shown in Figure 5.

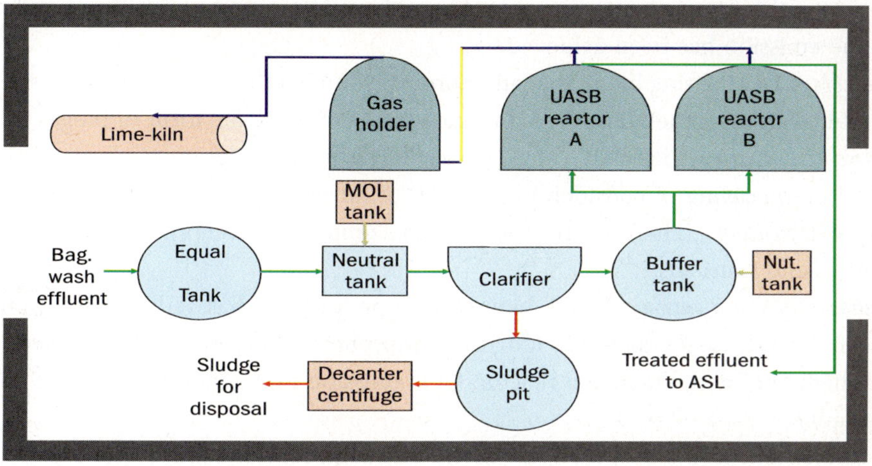

Figure 5 Process flow in a biomethanation plant

The lime kiln is designed to produce 170 MT of burnt lime per day. Furnace oil, a fossil fuel, was used as fuel, which used to release CO_2. Now, the biogas generated from biomethanation process has partially replaced the furnace oil. The consumption of furnace oil used to be in the range 180–195 litres/tonne of lime, which has now reduced to 115–125 litres/tonne of lime.

Table 1 Biogas plant performance

Year	*Waste water treated (million m³)*	*COD reduced (MT)*	*Biogas produced (million m³)*	*Furnace oil saving (kl)*	*GHG reduction CO_2eq (MT)*
2003	14.16	4 716	2.37	1 450	16 965
2004	29.47	8 686	3.51	2 009	29 579
2005	31.46	11 036	4.75	2 849	38 103
2006	29.62	10 144	5.02	3 013	35 500
2007*	30.11	10 485	4.77	2 864	35 554
Total	134.82	45 067	20.42	12 185	155 701

*Till September 2007
MT – million tonnes; GHG – greenhouse gas; COD – chemical oxygen demand

Cost–benefit analysis

Cost of biomethanation plant:	Rs 44 million
Cost of furnace oil at the time of inception:	Rs 10/litre
Annual savings:	Rs 26.2 million
Operating cost of the plant:	Rs 15.9 million
Net saving per year:	Rs 10.3 million

The cost of the furnace oil has almost doubled now, which has reduced the payback period and increased the internal rate of return.

Benefits

- *Pollution control* Treating effluent in anaerobic lagoons used to release methane, a potent GHG, into the atmosphere. By using biomethanation process, significant reduction of methane emission has been achieved.
- *Renewable energy* Using biomethanation process to treat effluent from the bagasse washing has resulted in the generation of clean renewable energy in the form of biogas.
- *Resource conservation* Partially replacing the furnace oil in the lime kiln with biogas has resulted in the saving of considerable amount of furnace oil, a fossil fuel.
- *Energy conservation* There is also reduced consumption of energy and consumables like chemicals and nutrients.
- *Greenhouse gas mitigation* Using biomethanation process has led to GHG mitigation. There is no release of methane into the atmosphere after replacing anaerobic lagoons with a high-rate biomethanation reactor. CO_2 emission in the lime kiln has also decreased remarkably.
- *Decreased organic load* The high-rate biomethanation reactor reduces the organic load on the downstream activated sludge process.
- *Savings* Net savings are envisaged to be to the tune of Rs 10.3 million per annum.

CASE STUDY AT A GLANCE

- The TNPL (Tamil Nadu Newsprint and Paper) was established by the Government of Tamil Nadu to produce newsprint, and printing and writing paper using bagasse.
- Treating bagasse wash waste water in anaerobic lagoons used to release methane, a potent GHG (greenhouse gas). To address this problem, a high-rate biomethanation plant replaced anaerobic lagoons, which is essentially an anaerobic reactor.
- Using biomethanation process to treat effluent from the bagasse washing has resulted in the generation of clean renewable energy in the form of biogas.
- Partially replacing the furnace oil in the lime kiln with biogas has resulted in the saving of considerable amount of furnace oil. This has also reduced consumption of energy and consumables like chemicals and nutrients.
- Using biomethanation process has led to GHG mitigation.
- Net savings per year at the time of installation of the biomethanation plant were envisaged to be Rs 10.3 million.

CASE STUDY 25 Aban Power Company Ltd

Figure 1
Aban Power Company Ltd

About the company

The Aban Power Company Ltd, set up at a cost of Rs 4300 million, is a 120-MW (megawatt) combined cycle power plant located in Tamil Nadu (Figure 1). The installed generation capacity of the plant is achieved through a 68.6-MW-capacity gas turbine and a 51.2-MW-capacity steam turbine. The grid requirements decide the operation pattern of the facility. Primarily, the plant is a base load operated unit. It uses natural gas as a fuel, which is supplied by GAIL (Gas Authority of India Ltd). The gas is not stored and is drawn directly through pipeline and is directly consumed after pretreatment.

Environmental profile

The company has an Environmental Policy at both corporate and facility levels. It also has an environment department. It is ISO 14001 and OHSAS 18001 certified.

Air pollution and its control measures

NO_x (oxides of nitrogen) are the main air pollutants emitted from the facility. DLN (dry low–NO_x) burners have been installed in the facility to control NO_x emissions (Figure 2).

Figure 2
DLN burner

Water pollution and its control measures

The main water pollutants present in the effluent generated from the facility are BOD (biochemical oxygen demand), TDS (total dissolved solids), and TSS (total suspended solids). BOD is controlled by treating the effluent in the sewage treatment plant based on the upflow anaerobic digester. The treated effluent is reused for watering the green belt inside the factory premises.

To treat TSS and TDS, the effluent is treated in the reverse osmosis plant, and the treated effluent is recycled and reused. The rejects are diverted to solar evaporation ponds. Thus, there is zero discharge of effluents.

Hazardous waste

The only hazardous waste generated is used oil. It is collected, stored, and disposed of to the pollution control board authorized recyclers.

Case study

Energy and resource conservation projects

Aban has taken many proactive measures to ensure that its activities do not have adverse impacts on the environment. Towards this end, it has taken up many projects involving process improvement or modification and installation of latest energy saving equipment to save natural resources like water, mitigate pollution, and conserve energy.

During the operational stage

In compliance with requirements (process improvement/modification)

Improved performance of the reverse osmosis plant

The presence of carbonates in the feed effluent used to result in the scaling of reverse osmosis membrane, leading to frequent cleaning. To address this problem, the acid addition treatment programme was implemented in 2007/08 and the feed pH was brought down from 7.0–7.5 to 6.0–6.2. Thus, the Langlier Saturation Index was brought close to zero or a slightly negative value. This resulted in the increase in the permeate by 0.9 m^3/h (Figure 3).

Figure 3
Reverse osmosis plant

Cost–benefit analysis

Annual water saving in 2006/07 was 2448 m^3, equivalent to Rs 70 000, and in 2007/08, 3366 m^3, equivalent to Rs 101 000. Investment made in 2006/07 was Rs 5000 and in 2007/08, Rs 25 000 (due to the increase in the cost of chemicals).

Beyond compliance to statutory requirements (proactive practices)

Process modification in the auxiliary cooling water system of the steam turbine

The cycle of concentration was increased from 5 to 8 in the auxiliary cooling water system of the steam turbine. This reduced effluent generation, chemical consumption, and load of effluent treatment.

Cost–benefit analysis

Investment was nil. The consumption of water reduced by 270 m^3/month. Savings accrued due to the reduction in chemical treatment was to the tune of Rs 6044 per month. Annual water savings are of the order of 6480 m^3, which is equivalent to savings worth Rs 345 000.

Construction and maintenance of a storm water reservoir inside the plant premises

The storm water reservoir of 18000 kilolitres capacity comprises nine percolation pits. The water is utilized for maintaining 14985 trees, 22000 shrubs, 11000 hedges, and 15000 m^2 of lawn. This reservoir also sustains groundwater in and around the plant area (Figure 4).

Figure 4
Storm water reservoir

Cost–benefit analysis

Annual water savings are to the tune of 36000 m^3, which is equivalent to savings worth Rs 1.08 million. Investment made in this project was Rs 5.4 million.

Energy saving measures through the implementation of environment management programmes under ISO 14001:2004 systems

Reduction of run hours of air-cooled heat exchanger fans of closed cooling water system

Objective

The main aim was to reduce the auxiliary power consumption by switching off the ACHE (air-cooled heat exchanger) fans that are used to cool the closed cooling water, which dissipates the heat absorbed from GTG, SWAS, and BFP LO cooler. Run hours of the ACHE fans were reduced by 3%. Thus, an average of 40% of the installed capacity is reduced (Figure 5).

Cost–benefit analysis

The cost of energy saved was Rs 2.06 million for the period of implementation, that is, from May 2006 to March 2008.

Figure 5
Air-cooled heat exchanger

Reduction of run hours of air-cooled condenser fans for steam turbine

Objective

The objective was to reduce auxiliary power consumption by switching off the ACC (air-cooled condenser) fans that are being used to convert ST exhaust steam to condensate, after work done in the steam turbine by air cooling method. Run hours of ACC fans were reduced by 2% (Figure 6).

Figure 6
Air-cooled condenser

Cost–benefit analysis

The cost of energy saved was Rs 1.13 million for the period of implementation, that is, from May 2006 to March 2008.

Reducing plant illumination power consumption

Case I

Existing condition

There are two lighting transformers feeding main lighting distribution board through switch fuse units. The incoming power to these lighting transformers is fed from the PCC 'A' bus and PCC 'B' bus by using switch fuse unit arrangement. Presently, only one transformer is used for supplying power to the main lighting distribution board by using bus coupler switch closed. The other transformer is in ideally charged condition. The no load is assumed to be 0.9% of the installed capacity, that is, 1.08 kW. There is no auto changeover facility for these transformers. The changeover operation is done manually. This reduction in the plant illumination power conservation has led to the conservation of natural resources like gas, minimization of pollution, and reduction in auxiliary power consumption.

Proposed condition (energy conservation)

- By keeping the other transformer in off mode, about 1 kW/hour of energy can be saved.
- Because of the heat dissipation by the transformer through no load losses, the switch gear hall temperature will also be minimized.
- There is no deviation in the operation procedure while keeping this transformer switched off.
- Changeover schedule is formulated for the transformers to be changed over once a week.
- Transformers are safe in off mode as they are air cooled.

Cost–benefit analysis

Auxiliary consumption was reduced by 1.08 kW/hour, or 9461 kW/year. In monetary terms, this amounts to a saving of Rs 20 814 per annum.

Case II

Existing condition

It was observed that the voltage supply level for the plant lighting is more. The lighting transformer output power to the main lighting distribution board was also found to be more. It was proposed to change the tap positions in transformers. After changing the tap positions, voltage and current are reduced. A total of 8.15 kW of power saving is achieved.

Cost–benefit analysis

Auxiliary consumption has reduced by 8.15 kW/hour, or 35 697 kW/year. This has led to an annual saving of Rs 78 533.

Case III

Existing condition

The steam turbine has 14 ACCs for condensing the exhaust steam generated from the steam turbine. The exhaust steam is being collected in a hot well. From here, the condensate is pumped to the HRSG (heat recovery steam generator) through a condensate extraction pump. The water is again converted into steam and used in the steam turbine for power generation. The water loss due to de-aeration, CBD, drain traps, and steam vents is the only make up required, which is being met by DM (de-mineralized) water. The make up is being done in the hot well tank through hot well make–up pump (DM transfer pump). The DM transfer pump is run continuously for 24 hours, and the control valve opens and closes according to the level and water requirement.

Test results

Two tests were carried out.

Observations from the first test

- 10% level dropping from 63.63% to 53.6% requires three hours.
- Same 10% level increase requires one hour.
- The pump can be switched off for 18 hours and can be run for six hours in six starts and six stops.

Observations from the second test

- 22% level dropping from 71.33% to 49.22% requires 5.30 hours.
- Same 22% level increase requires 2.10 hours.
- The pump can be switched off for 17.22 hours and can be run for 6.48 hours in three starts and three stops.

Combined result

- As an average of the above two tests, the pump can be stopped for 17.5 hours and operated for 6.5 hours when HRSG blowdown is 1 pulse.
- The pump can be stopped for 14 hours and operated for 10 hours when HRSG blowdown is 2 pulse.
- So, in a month of 30 days, the HRSG blowdown of pulse 1 should be given for 23 days and that of 2 pulse for seven days. Total run hours in a year are 2519.5 hours, and total stop hours are 5736.5 hours. Motor rating is 2.2 kW.

Cost–benefit analysis

Assuming that the motor is delivering an average of 2 kW of power, total power saving achieved is 11 473 kW, which gives a cost benefit of Rs 25 240.6/year. Lube oil consumption and maintenance cost are reduced.

Modifications required

- The pump is to be configured in the DCS (distributed control system) to start when the level is less and stop when the level is more.
- Audible alarms to be generated in the DCS for indicating the starting and stopping levels.

During the project stage

Beyond compliance to statutory requirements (proactive practices)

Using dry low–NO_x burners in the gas turbine

Natural gas is used as a fuel in the Aban facility, which emits NO_x. Conventionally, NO_x emissions are controlled by using pure DM water or injecting steam along with the fuel. This method involves consumption of natural resources like water, power, additional fuel, and chemicals, and, in turn, needs pollution control measures.

The alternative latest technology for controlling NO_x emissions is using DLN burners, which do not use DM water or steam and maintain the NO_x levels at less than 20 PPM (parts per million).

Cost–benefit analysis

Investment made in installing DLN burners was Rs 160 million, and the payback period due to the savings in DM water, steam, fuel, and chemical dosing would be one year.

Using air-cooled condensers in place of conventional cooling towers in the steam turbine

A typical power plant using cooling towers to condense the steam generating from the steam turbine exhaust will require 2500 litres/MWh of water. Aban has installed ACCs in place of conventional cooling towers, which consume 102 litres/MWh of water. Thus, there is a saving of 2398 litres/MWh. Overall, there is reduction in water consumption at the source, chemical consumption, and pollution. All this ensures that there is zero discharge.

Cost–benefit analysis

Annual water saving in 2005/06 was 1 353 056 m^3 (Rs 40.6 million); in 2006/07, 2 061 368 m^3 (Rs 61.8 million); and in 2007/08, 2 142 673 m^3 (Rs 64.2 million). Investment made was Rs 150 million.

Using air-cooled heat excchangers in place of conventional cooling towers in the gas turbine

A power plant having gas turbine (68.6 MW rated generation) and auxiliaries using cooling towers will require 6.07 kilolitres/hour of water for make up (based on existing set-up). Aban has installed ACHEs (closed cooling water system) in place of conventional cooling towers. This has brought down the water requirement to nil. Thus, there is a saving of 6.07 kilolitres/hour of water. Overall, there is reduced water consumption at the source, chemical use, and pollution, and all this makes the company is a zero discharge facility.

Cost–benefit analysis

Annual water saving in 2005/06 was 36 000 m^3 (Rs 1.08 million); in 2006/07, 51 000 m^3 (Rs 1.53 million); and in 2007/08, 51 000 m^3 (Rs 1.53 million). Initial investment made was Rs 9 million. Payback period is another 3.5 years.

Benefits

The projects undertaken by the Aban facility have resulted in a significant reduction in raw water consumption and NO_x emissions (Figures 7 and 8), along with energy conservation.

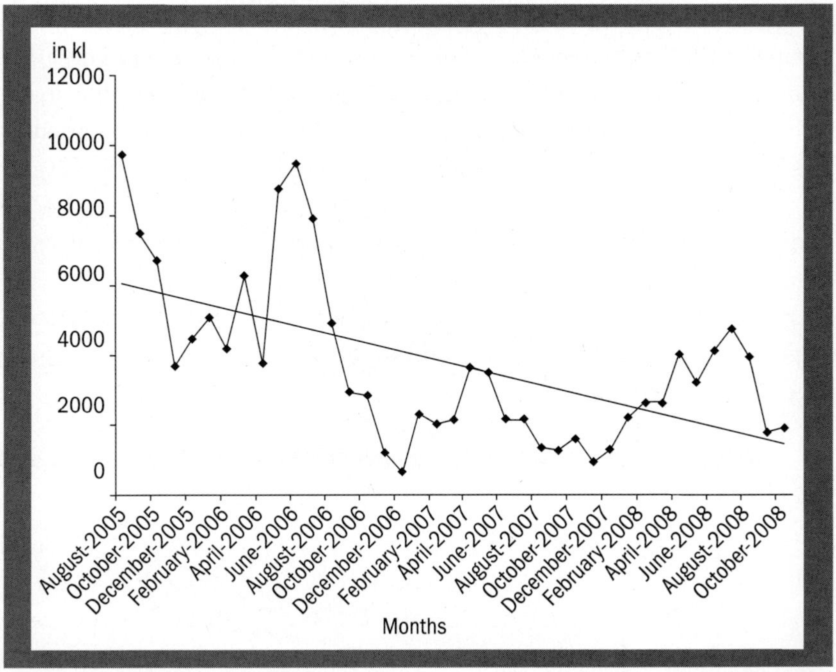

Figure 7
Reduction in raw water consumption

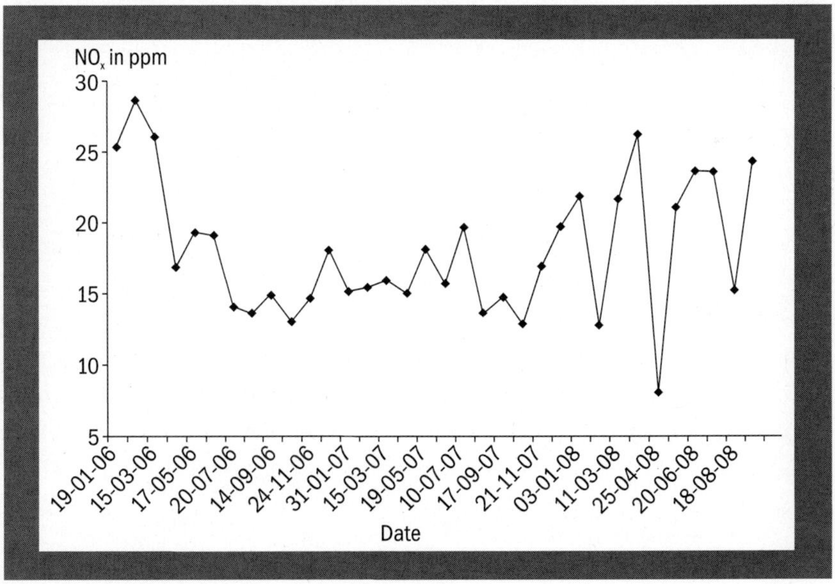

Figure 8
Reduction in NO_x levels in the main stack (standard 75 PPM)

CASE STUDY AT A GLANCE

- The Aban Power Company Ltd is a 120-MW (megawatt) combined cycle power plant using natural gas as fuel.
- The company has implemented various projects for saving water. These include improving reverse osmosis performance, increasing cycle of concentration in cooling towers, constructing of storm water reservoirs, and using air cooled condensers.
- With the dual aim of conserving water and reducing the emissions of NO_x (oxides of nitrogen), the company has installed DLN (dry low–NO_x) burners.
- The company has taken various measures towards conserving natural resources and saving energy.
- The proactive initiatives taken by the company have, overall, reduced water consumption at the source, chemical consumption, and pollution, and ensure zero discharge.

CASE STUDY 26 Hindustan Zinc Ltd

Figure 1
Chanderiya Lead Zinc Smelter, Hindustan Zinc Ltd

About the company

The HZL (Hindustan Zinc Ltd), a company of Vedanta Resources Group (London, UK), is India's only integrated producer of zinc and lead and is among the world's leading integrated zinc and lead producers (Figure 1). It produces 426 323 MT (million tonnes)/annum of zinc and 63 556 MT/annum of lead. The major by-products produced are silver (80.4 MT/annum), sulphuric acid (753 966 MT/annum), and cadmium metal (537 MT/annum). Its mining and smelting operations are mainly located in the states of Rajasthan and Andhra Pradesh. The smelters are situated at Chanderiya, Debari, and Visakhapatnam, and the mines are situated at Zawar, Dariba, and Rampura Agucha.

The turnover of the company in 2007/08 was Rs 78 780 million.

Environmental profile

The HZL strongly believes in sustainable development. To ensure that all its operations are environment friendly and resource efficient, the company has in place a combined Health, Safety and Environment Department. It also has an Integrated Management System, which is ISO 9001, ISO 14001, OHSAS 18001, and SA 8000 certified.

Under its environment initiatives, the company has adopted various water conservation and energy conservation measures. It has also switched over to renewable sources of energy and has managed to reduce its carbon footprint significantly.

Air pollution

The facility mainly emits SPM (suspended particulate matter), RSPM (respirable suspended particulate matter), SO_2 (sulphur dioxide), and fugitive pollutants during its operation. Tall stacks have been constructed in the plants to ensure better atmospheric dispersal of the gases released through them.

Air pollution control measures

SPM

Bag filters have been installed in stacks to control SPM.

SO_2

- In the acid plant attached to the pyro smelter, double conversion double absorption method was used earlier, which used to result in the release of 2.7 kg SO_2 per tonne of acid. New and improved technologies were implemented to address this problem, and many changes were made in the acid plant, including revamping and replacing major equipment. Because of these changes, the SO_2 to SO_3 conversion efficiency has enhanced and SO_3 absorption system has improved. Also, now less than 2 kg SO_2 per tonne of acid is produced.
- A TGT (tall gas treatment) plant is in operation in the acid plant, which treats the outgoing gases and ensures that the emission levels of SO_2 and acid mist are below the prescribed standards.
- An online SO_2 analyser has been installed in the TGT stack to monitor and control SO_2 emissions.

RSPM and fugitive emissions

RSPM and fugitive emissions are controlled by electrostatic precipitators, scrubbers, cyclones, and bag filters that are installed in all polluting units. Dust is contained by spraying water, washing roads, and mechanized vacuum sweeping.

Water pollution control measures

The facility has 'zero discharge'. The CLZS (Chanderiya Lead Zinc Smelter) Complex has two ETPs (effluent treatment plants; capacities 400 m^3/h and 275 m^3/h) to treat the effluent generated from the pyro smelter, Ausmelt lead plant, and hydro smelters. About 6000 m^3/day of the treated effluent is further treated in the RO (reverse osmosis) plant for polishing (Figure 2). The treated water is recycled for washing roads, watering the green belt developed in the premises of the facility or is used as process water, thus ensuring zero discharge.

Hazardous waste

The facility generates various types of hazardous waste. Select waste types, along with their disposal methods, are listed below.

- *Spent catalyst* Stabilized with lime and cement before being disposed of in SLFs (secured landfills).
- *Jarosite (DWT)* Stabilized with cement and lime as 'Jarofix' and disposed of in designated Jarofix Yard.

Figure 2
Reverse osmosis plant

- *Cobalt cake* Disposed of in the SLF after stabilization.
- *ISF (imperial smelting furnace) slag* Stacked in the slag yard. Approval has been sought from the RSPCB (Rajasthan State Pollution Control Board) for utilizing it for making cement, road, brick, and tile.
- *Antimony dross* Recycled in the plant and also sold to the CPCB (Central Pollution Control Board)-registered recyclers.
- *ETP sludge* Recycled in the sinter plant; disposed of in the SLF after stabilization.
- *Used/waste oil* Sold to registered recyclers.
- *Purification cake* Sold to the CPCB-registered recyclers.

Other waste

The CLZS has installed an OWC (organic waste converter) to convert canteen and kitchen food waste into organic manure (Figure 3). The OWC presently processes 125 kg of food waste per day, generating 40 kg/day of manure, which is applied to plants grown in the premises.

Case study

Producing sulphuric acid from lean and varying SO_2-bearing gases in the Ausmelt lead plant

Batch process and SO_2 generation

The HZL commissioned a lead smelter in its CLZS unit, Chittaorgarh, in 2005. The smelter follows top submerged lance technology, supplied by M/s Ausmelt Ltd, Australia, for smelting lead sulphide concentrates. Though the technology has an established continuous process for smelting, the HZL has opted for the batch process, in which smelting, slag reduction, and slag cleaning operations take place in a single furnace. In this process, SO_2 concentration in the furnace off-gas varies from 0.1% to 12% in a cycle time of 11 hours. So, the challenge that the smelter faced when it decided to adopt the batch process was to treat the off-gas in an efficient and environment-friendly manner.

Figure 3 Organic waste converter

Various options explored for treating SO_2

Many options for treating this gas were explored, as listed below, but none was found to be feasible.

- A stand-alone acid plant to treat SO_2 to produce sulphuric acid is not technically viable. During the slag reduction and slag cleaning stages, the off-gas contains very low concentration of SO_2 (approximately 0.2%), and the acid plant cannot run on such a dilute feed.
- Alternatively, the off-gas generated during smelting can be fed into the existing acid plant after modifying it, while the off-gas from the slag reduction and cleaning stages can be treated in a lime scrubber. This will require a capacity addition of 45% in the plant to accommodate the extra gas. But there is a risk of the operation of the sulphuric acid plant getting hampered due to the gas flow being intermittent.
- Alkali scrubbing of SO_2 generates huge amounts of solid waste, giving rise to the problem of its disposal. The operating costs are also very high.
- Ammonium sulphate production and zinc calcine scrubbing are also not commercially viable.

Process followed in the facility for recovering SO_2

An innovative approach (Cansolv technology) was adopted by the company, wherein SO_2 is removed in the absorption tower by its absorption in an aqueous amine solution. This solution containing SO_2 is collected in a buffer tank and then pumped to a disorption tower. Downstream of the pumps, the liquid is heated in plate heat exchangers and then in a reboiler. Heating strips the SO_2 from the hot liquid. Water-saturated SO_2 leaves the desorber. Downstream of the desorber, SO_2 gas is cooled and water vapour is condensed. By using buffer tanks, the fluctuations of the SO_2 from the Ausmelt furnace are changed into a continuous flow of SO_2-containing liquid and then to a flow of concentrated SO_2 gas. Figure 4 gives the process detail.

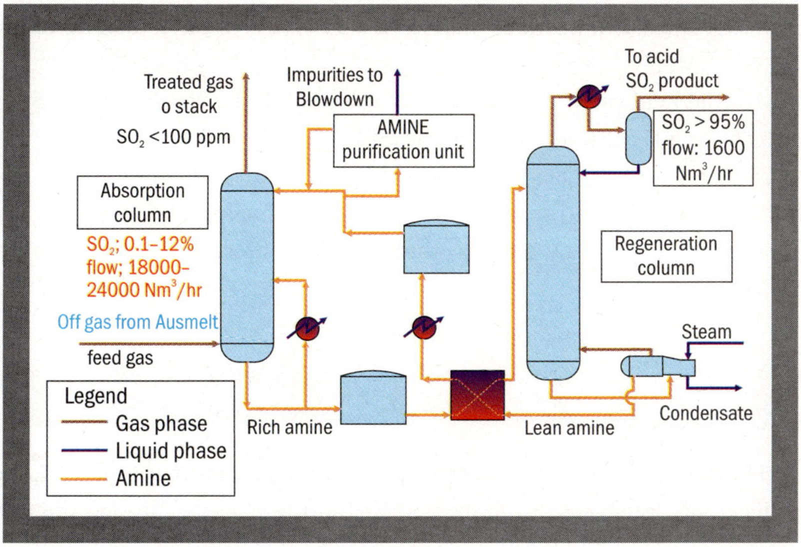

Figure 4
Process diagram of Cansolv technology

Use of the recovered gas in the acid plant

The SO_2 thus recovered can be used for increasing the strength of the SO_2 feed gas of the existing contact acid plant to produce sulphuric acid. A small modification was done in the acid plant to accommodate this extra SO_2. Using this process, the facility was able to address the problem of low SO_2 concentration in the existing acid plant based on sinter plant gases.

Salient features of the process

This regenerative-type SO_2 recovery system has following features.

- This process is totally regenerative-type SO_2 recovery system, and SO_2 concentration in the off-gas is less than 100 PPM (parts per million).
- This process can handle a varied range of SO_2 concentration in its inlet off-gas. In the CLZS, SO_2 concentration varies from 12% to 0.1%. In its outlet, the process produces a continuous flow of concentrated SO_2 gas (95% SO_2).
- This 95% SO_2 gas is used in the existing acid plant to produce sulphuric acid.
- The efficiency of the acid plant, which operated on low concentration SO_2 gas, has increased with the availability of concentrated SO_2.
- The acid plant has stopped using the pre-heater due to the availability of high concentration SO_2, thus saving on fuel.
- This process is very cost-effective when compared to other treatment processes.

Cost

The HZL invested more than Rs 500 million to implement this process of SO_2 recovery.

Benefits

- The problem of SO_2 emissions has been addressed to a large extent, as the process not only recovers SO_2 but also puts it to beneficial use.

- The recovered SO_2 is used to produce sulphuric acid (150 MT/day), leading to increase in revenue.
- The efficiency of the acid plant, which used to operate on low concentration SO_2 gas, has increased with the availability of concentrated SO_2.
- Enhanced performance of the acid plant has resulted in the reduction of SO_2 emission– from 4 kg/MT of acid produced to less than 1.5 kg/MT of acid produced.
- As the pre-heater is not used any more in the acid plant, there is a considerable saving in the fuel oil.

CASE STUDY AT A GLANCE

- The HZL (Hindustan Zinc Ltd), a company of Vedanta Resources Group (London, UK), is India's only integrated producer of zinc and lead and is among the world's leading integrated zinc and lead producers.
- The turnover of the company in 2007/08 was Rs 78 780 million.
- The CLZS (Chanderiya Lead Zinc Smelter) of the HZL adopted a unique approach for recovering SO_2 (sulphur dioxide) from the off-gas generated during its smelting operations.
- The recovered concentrated SO_2 is used in the sulphuric acid plant to produce sulphuric acid.
- With the adoption of this process, the problem of SO_2 emissions has been addressed to a large extent.
- The efficiency of the acid plant, which used to operate on low concentration SO_2 gas, has increased with the availability of concentrated SO_2.
- As the pre-heater is not used anymore in the acid plant, there is a considerable saving in the fuel oil.

SECTION

IV

Environment management system

The Planning Commission (2006)[1] observes that the private sector, including the corporate sector, accounts for 70% of the total investment in the economy, and it emphasizes the critical role of this sector in achieving the objective of faster and more inclusive growth in India. The World Bank (2007),[2] on the other hand, recognizes the existence of wide-ranging environmental risks and problems in the country and observes that environmental sustainability might be the next greatest challenge in the country's development path. To deal with these impacts, India has developed a comprehensive set of environmental laws and institutions. India is amongst the first countries to have provided for the protection and improvement of the environment in its Constitution. Recognition of the need for environmental protection in India dates back to as early as the 1970s. India's participation in the multilateral environment agreements has also helped enhance the domestic environmental legislative framework. Moreover, the government has issued guidelines for production process certifications to ensure that environmentally sound production and management practices are adopted in India.

New environmental norms and regulations introduced over the past few decades have created a demand for environmental services, especially those directly related to pollution control in the manufacturing sector.[3] Further, the World Bank (2007) observes that the rising public demand for better environmental quality is being increasingly matched by voluntary environmental performance obligations from India's large-scale corporate sector and industry, asserting a prominent role of the country in the global market. In this regard, the MoEF (Ministry of Environment and Forests) launched the Charter on 'CREP (Corporate Responsibility for Environmental Protection)'[4] in March 2003, with the purpose

---

1 Planning Commission. 2006. *Towards faster and more inclusive growth.* An Approach to the Eleventh Five-year Plan.

2 World Bank. 2007. *Strengthening institutions for sustainable growth.* Country environment analysis for India. South Asia Sustainable Development Unit, World Bank, New Delhi, India.

3 Chanda R (ed.). 2006. *Trade in services and India.* Prospects and strategies. CENTAD and Wiley India Pvt. Ltd, New Delhi, India.

4 Website of Charter on Corporate Responsibility for Environmental Protection <http://www.cpcb.nic.in/crep.php>, last accessed on 13 May 2009

to go beyond the compliance of regulatory norms for the prevention and control of pollution through various measures, including waste minimization, in-plant process control, and adoption of clean technologies. The Charter has set targets concerning conservation of water and energy, recovery of chemicals, reduction in pollution, elimination of toxic pollutants, and management of residues that are required to be disposed of in an environmentally sound manner.

It is perceived that the corporate response to the environment is either based on environmental risks and market-based opportunities or environmental pressures like legislation and constraints within the firm. A competitive advantage can be achieved not merely by keeping abreast of environmental developments, but also by initiating change within an organization and responding with new environment-friendly products and production processes. The ultimate aim of the corporate environmental management must be to reach a situation where companies operate in a way that is consistent with the concept of sustainable development. For businesses, the necessary steps are the development of environmental management systems that promote quality, a commitment by the management as well as employees to environmental issues, and to ensure that environmental performance is improved over a period of time.

Environmental management systems for companies help them in not only meeting the legislative requirements but also implementing, managing, and fostering voluntary initiatives at the corporate level. The management systems cover not only management responsibilities but also the responsibility and tasks of every individual in an organization.

CASE STUDY 27 Gujarat Co-operative Milk Marketing Federation Ltd

About the company

The GCMMF (Gujarat Co-operative Milk Marketing Federation Ltd) is India's largest food-producing company with an annual turnover of more than Rs 53 billion ($1.3 billion), which makes it a billion dollar entity. Its daily milk procurement is more than 7.5 million kg from 13 141 village milk cooperative societies, 13 member unions covering 20 districts, and 2.7 million milk producers. With an aggregate milk processing capacity of 10.16 million litres per day, it is Asia's biggest dairy business venture. It is the exclusive marketer of 'Amul' and 'Sagar' brand products. Its product range comprises milk, milk powder, health beverages, ghee, butter, cheese, ice-cream, chocolates, and traditional Indian sweets.

The GCMMF cannot be viewed only as a business organization. It is an institution created by milk producers to primarily safeguard their interests economically, socially, as well as democratically. While business houses distribute profits to shareholders, in the GCMMF, profits are given back to farmers. For every rupee earned by the GCMMF, roughly 75 paise flows back to farmers.

Environmental profile

All the products of the GCMMF are manufactured under the most hygienic conditions. All dairy plants of the unions are ISO 9001 and HACCP certified, which ensures the best quality of products all the time.

The company's TQM (Total Quality Management) ensures high quality of products. Additionally, since the company is a marketing organization, it does not generate any pollutant.

Case study

Green Gujarat Tree Plantation Campaign by milk producers of dairy cooperatives

Background

The state of Gujarat has experienced a great degree of forest loss over the years. A recent report by the Forest Survey of India indicates that in the last two years, Gujarat has lost close to 100 km^2 of its forested area due to anthropogenic causes, which is a good 13% of the national loss. The total forest cover in Gujarat is only 11.4% of the state's geographical area.

Considering these dismal statistics, milk producers in Gujarat initiated a tree plantation programme that was aimed at increasing the green cover of the state.

Initiative

On the occasion of the country's 60th Independence anniversary (15 August 2007), more than 2.5 million milk producers of 13 member unions of the GCMMF, across 20 districts of Gujarat, came together to launch a tree plantation drive in the state of Gujarat.

The idea of the plantation drive took shape when the state-level apex body of dairy farmers in Gujarat, the GCMMF, floated a novel concept of 'one member one tree plantation'. To put this concept into practice, a design team comprising representatives of member unions was formed. The team, in turn, involved the farmer members of village dairy cooperative societies.

To implement this programme, the design team charted a road map for various activities. Execution teams were formed at district union level to put the plan into action. Village-level coordinators were identified and trained on the activity of tree plantation. A massive awareness generation drive on the benefits of tree plantation was launched among farmer members.

The execution team coordinated with nurseries to supply the farmer members with tree saplings of their choice. Logistics for transporting saplings from nurseries to villages were provided by the GCMMF.

On 15 August 2007, each member took an oath to plant a sapling. Altogether, 1.89 million tree saplings were planted across 19 districts of Gujarat. Of these, 57% (1.07 million saplings) tree saplings survived, providing about 10 160 hectares of additional tree cover, thus adding 0.47% of tree cover in the state. With this successful culmination of the first phase of the plantation drive, the Village Dairy Cooperative Societies of Gujarat decided to commemorate 15 August as 'Green Revolution Day by Afforestation'.

Inspired by the success of 2007, a more ambitious target was kept for 2008. On 15 August 2008, milk producers of 7071 village dairy cooperative societies of 12 member unions of the GCMMF planted approximately 5.14 million tree saplings across 21 districts in the state. The tree plantation programme in 2008 was upgraded to 'three trees per member'. Considering that the survival rate of the saplings in 2008 would remain the same as that in 2007, 2.93 million trees are expected to survive, providing an additional 27 660 hectares of tree cover, thus adding 1.27% of tree cover in the state. Thus, 2.5 million milk producers of Gujarat planted more than 7 million saplings in just two years. Considering the survival rate of 57%, 1.74% of forest cover would be added to the existing green cover in Gujarat.

The members of the GCMMF have, thus, set an example for all other institutions by initiating a programme to 'turn India green', which is especially relevant in an era of global warming and deepening environmental crisis.

Cost–benefit analysis

No direct cost was incurred for conducting the programme. Saplings were obtained from government nurseries free of cost, with a few exceptions. The saplings were transported from nurseries to the plantation site using the existing channel of milk transportation at milk union level, incurring little additional cost. In some places, village dairy cooperative societies had to make their own arrangement to procure and transport saplings. So, it is estimated that a cost of 50 paise was incurred per sapling. Thus, for planting 7.1 million saplings in two years, approximately, Rs 3.5 million was spent in total.

It has been estimated that when one tree is cut, there is a loss of Rs 3.3 million in the form of lost ecosystem services. So, considering a net benefit of Rs 3.3 million from a single tree, the total net benefit in two years will be Rs 128.7 million.

Benefits

Milk producers of member unions of the GCMMF have planted more than 7 million trees in 2007 and 2008, which are expected to increase the land under green cover in the state by 37 821 hectares, adding 1.74% of forest cover in the state. Increased forest cover would result in better environmental conditions, improved landscape, stable climate, and protection and conservation of wildlife.

CASE STUDY AT A GLANCE

- The GCMMF (Gujarat Co-operative Milk Marketing Federation Ltd) is India's largest food-producing company with annual turnover of more than Rs 53 billion ($1.3 billion), making it a billion dollar entity.
- It is the exclusive marketer of 'Amul' and 'Sagar' brand products. Its product range comprises milk, milk powder, health beverages, ghee, butter, cheese, ice-cream, chocolates, and traditional Indian sweets.
- The company undertook an ambitious tree plantation programme in two phases in a bid to increase the green cover of the state.
- The initiative saw the successful plantation of 1.89 million tree saplings in the first phase of the tree plantation programme held on 15 August 2007. Of these, 57% (1.07 million saplings) tree saplings survived, providing about 10 160 hectares of additional tree cover, thus adding 0.47% of tree cover in the state.
- In the second phase of the programme held on 15 August 2008, 5.14 million tree saplings were planted under the 'three trees per member' drive. Approximately, 5.14 million tree saplings across 21 districts in the state were planted. Considering that the survival rate of the saplings in 2008 would remain the same as that in 2007, 2.93 million trees are expected to survive, providing an additional 27 660 hectares of tree cover, thus adding 1.27% of tree cover in the state.
- Altogether, more than 7 million trees were planted in 2007 and 2008, which are expected to increase the land under green cover in the state by 37 821 hectares, adding 1.74% of forest cover in the state.

CASE STUDY 28

Sesa Goa Ltd

Figure1
A view of the Codli Mine

About the company

Sesa Goa is an associate of Vedanta, an FTSE 100 metals and mining company, with its principal operations in India, Zambia, and Australia. For the past five decades, Sesa Goa has been involved in metal ore mining, beneficiation, and exports. It has also diversified into the manufacture of pig iron and metallurgical coke. The major metals it produces are aluminium, copper, zinc, lead, and iron ore.

Sesa Goa Ltd, the flagship company of the Sesa Group, is India's largest private sector exporter of iron ore, with mining operations in Goa, Karnataka, and Orissa. It is engaged in iron ore prospecting, mining, processing, and exporting. It is also actively involved in support activities such as shipping and shipbuilding.

Sesa Goa's main mining operations are at Codli and Sonshi. The mine at Codli is a multi pit operation (Figure 1).

Environmental profile

Mining is generally perceived as having adverse environmental impacts on the environment. Sesa Goa Ltd has assiduously incorporated environment-friendly practices in all its operations. Its Environmental Policy places emphasis on optimum utilization of natural resources, rehabilitation of the mine in a sustainable manner, and preventing pollution and continually improving the system. The facility has an in-house environmental laboratory for monitoring environmental parameters.

The company's Environment Management System is ISO 14001: 2004 certified at both corporate and facility levels.

Air pollution and its control measures

The mining operations carried out by the facility generate SPM (suspended particulate matter) and RSPM (respirable suspended particulate matter). The company has taken several measures to mitigate these pollutants. Drilling is generally done during monsoon season. During other seasons, wet drilling method is employed. Moisture content of the ore and associated waste is usually high, which prevents the dust from being airborne. Water is sprayed frequently on haul roads, and dust extractors are used to keep the roads dust free. Trucks carrying ore are covered and are made to pass through upgraded wheel wash system before leaving the mining area. Dust suppression systems are provided in all the units. Ultra fines are stored in specially constructed shades and are transferred with the help of closed conveyor belts.

Noise pollution and its control measures

Noise is monitored periodically using noise- monitors. The machines used for drilling are subjected to regular maintenance and kept in good working conditions. Acoustic enclosures have been provided for all generator sets. A green belt has been developed around the mining area, which attenuates the noise. Personnel working in high noise areas use earplugs.

Water pollution and its control measures

TSS (total suspended solid) is the main water pollutant. It is controlled by managing waste dumps efficiently. All the active dumps are located outside the mining area. These dumps are terraced. Water is channelized through garland drains into settling ponds located at the toe of the dumps, where it is treated with lime, if required. Finally, it is discharged from a common discharge point, which is monitored regularly.

During rainy season, preventive measures are taken to minimize dump run-off. Geotextiles are used and Styllo grass is grown on the dump surface to prevent soil erosion and ensure that TSS level in the dump run-off is within the prescribed standards.

Hazardous and other waste

Used oil and oil-soaked cotton are the hazardous wastes generated from the facility. The used oil is sold to the CPCB (Central Pollution Control Board)-authorized recyclers, and the oil-soaked cotton is incinerated. Apart from these, metal scrap and rubbish scrap are also generated, which are sold to recyclers.

Case study

Non-woven erosion control blankets, or geotextiles: a new approach for managing dumps

Overburden management is one of the major challenges for the mining industry based on the ore-to-waste ratio. The ore-to-waste ratio in the facility is 1:2.5. Handling such a huge quantity of waste generated during mining operations is fraught with environmental problems, especially in states like Goa, which receive 3000–4000 mm of rain annually during the rainy season spanning four months. The waste gets washed away in the run-off, contributing to TSS and leading to the siltation of rivers, rivulets or streams.

Initially, mining operations were carried out in the upper layers, and the mine spread was horizontal. So, there was abundant supply of laterite for covering the clayey portion of the mine dump and for making filter rock walls. This stabilized the dumps and checked the problem of erosion and siltation of rivers and rivulets from the run-off from the mining area.

However, as the mining went deeper, the reject became clayey in nature, which does not support plant growth and is prone to erosion. Stabilization of dumps and prevention of siltation posed a challenge at this point. To address this twin problem, erosion control blankets, or geotextiles, are used. Dumps are properly terraced and graded uniformly by removing boulders and stones, before the onset of monsoon. These are then layered with topsoil and soil conditioner. Geotextile mats are anchored on the surface of the dumps with 'U' hooks and pegs of green leaf manure plants (Figure 2). Dumps are further stabilized with a green cover. Grass and other plants like glyricidia are grown in between two geotextile mats (Figure 3). Grass species that control erosion, like *Styllosathus scabra*, are grown on the slope of the dumps.

Figure 2
Freshly laid geotextile mats

Geotextiles act as a shield and reduce the impacts of rain on the dump surface, thus preventing erosion. These also conserve moisture in the soil by acting as organic mulch, thus providing favourable conditions for the growth of vegetation on the otherwise infertile clay waste. These geotextiles decompose within three years, but by that time, dumps are already stabilized with vegetation.

Figure 3
Plants being grown on the geotextile mats

The geotextile mats have been used in specially designed filter beds at the toe of the dumps, along with layers of different sizes of pebbles and sand. This arrangement traps the fine particles from the dump run-off, thus reducing the TSS in the run-off.

The geotextiles are made of natural fibres like jute, coir, cotton, and so on. These are thin and permeable. Till date, the company has successfully used 250 000 m^2 geotextiles made of coir and 50 000 m^2 of geotextiles made of jute.

Table 1 indicates the reduction in TSS level achieved over the years by using geotextiles.

Table 1 Reduction in TSS (total suspended solids) (prescribed standard 100 mg/l) level over the years by using geotextiles

Year	*Area covered with geotextiles* (m^2)	*Average TSS* (mg/l)
2003	5 000	45.2
2004	12 000	11.5
2005	13 000	5.5
2006	15 000	14.8
2007	37 800	12.5
2008	35 600	21.83

Benefits

- *Reduced TSS level in the run-off* Geotextiles act as shields and lessen the impacts of rainfall on the waste, which otherwise would get washed away in the run-off, choking rivers and streams, thus checking erosion. These also trap fine particles in the filter beds at the toe of the dumps, reducing the TSS level in the dump run-off.

- *Stabilization of dumps* After they are covered with geotextiles, the dumps get a green cover in a period of three years, which stabilizes them.
- *Reclamation of the mine area* After mining is completed, the mine area is reclaimed effectively using this approach.

CASE STUDY AT A GLANCE

- Sesa Goa Ltd is India's largest private sector exporter of iron ore, with mining operations in Goa, Karnataka, and Orissa.
- It is engaged in iron ore prospecting, mining, processing, and exporting.
- The ore-to-waste ratio in the facility is 1:2.5.
- Handling such a huge quantity of waste generated during mining operations is fraught with environmental problems, especially in Goa, which receives 3000–4000 mm of rain annually during the rainy season spanning four months. The company adopted a novel approach for managing dumps by using non-woven erosion control blankets, or geotextiles.
- Geotextiles act as a shield and reduce the impacts of rain on the dump surface, thus preventing erosion.
- These also conserve moisture in the soil by acting as organic mulch, thus providing favourable conditions for the growth of vegetation on the otherwise infertile clay waste.
- The dumps get a green cover, which leads to their stabilization.
- The geotextiles also trap fine particles in the filter beds at the toe of the dumps, reducing the TSS (total suspended solid) level in the dump run-off.

CASE STUDY 29

Northern Coalfields Ltd

About the company

The NCL (Northern Coalfields Ltd), a subsidiary of the CIL (Coal India Ltd), was formed in 1986. Presently, the coal mining activities of the company are spread over an area of 100 km^2. The company produces thermal coal through 10 mechanized opencast mines located in Madhya Pradesh (accounting for 70% of coal produced) and Uttar Pradesh (accounting for 30% of coal produced). A steep rise is expected in the coal demand in the country, from 58.64 MT (million tonnes) in 2007/08 to 78.44 MT in 2011/12, to fulfil power and energy needs of the country. To meet this demand, the NCL has planned eight new projects, of which five have already been approved by the government. The turnover of the company in 2007/08 was Rs 54.55 billion.

Environmental profile

The company has an Environment Policy in place. Thrust areas identified by the company and included in the policy are as follows.

- Minimizing the impacts of mining operations on land, water, and air.
- Optimum utilization of resources in the areas of energy and water.
- Continual improvement in environment quality and prevention of pollution.
- Compliance with applicable environmental laws and regulations.
- Training of employees.

The company has a separate environment department at the corporate level and also an independent department of safety and health. Its EMS (Environmental Management System) manual is certified by M/s DNV (Det Norske Veritas, a consultant for ISO 14001) under ISO 14001: 2004. To achieve excellence in environmental performance and to meet the corporate goals, the EMS manual is followed diligently in all its facilities. Operations with potentials to cause adverse environmental impacts are also identified. The company takes utmost care to effectively integrate its existing system components with ISO 14001 standard's framework.

Air pollution

The main air pollutants emitted at the facilities are SPM (suspended particulate matter), RSPM (respirable suspended particulate matter), SO_2 (sulphur dioxide), and NO_x (oxides of nitrogen).

Air pollution control measures

- Spraying water by mobile sprinklers on haul roads and by fixed sprinklers at coal receiving pits, coal bunkers, transfer points, and loading points of the CHP (coal handling plant).

- Using dust extractors in the CHP and drilling equipment.
- Fully enclosing the CHP to prevent dust from going out of the CHP.
- Loading moist coal to MGR system through rapid loading system.
- Installing fire hydrant systems for the CHP and coal dumps.
- Black topping of service roads.
- Undertaking plantation.
- Providing dust masks and dust-proof cabins to HEMM (heavy earth moving machinery) operators.
- Undertaking routine maintenance and periodic overhauling of HEMM.

Noise pollution control measures

- Blasting operations are carried out between 12 noon and 3.00 p.m. only, that is, during shift change.
- Ear muffs and ear plugs are provided to employees, wherever required.
- Extensive plantation has been undertaken to attenuate noise.

Water pollution

The main water pollutants found in the waste water are BOD (biochemical oxygen demand), COD (chemical oxygen demand), pH, oil and grease, TSS (total suspended solids), TDS (total dissolved solids), and heavy metals.

Water pollution control measures

- ETPs (effluent treatment plants) have been installed in the down flow line of the workshop as well as mine discharges.
- Silt arrestors/silt ponds/sedimentation ponds have been constructed garlanding the OB (overburden) dumps to arrest the flow of silt to the river/waterbody.
- Oil and grease traps are installed at all workshops and ETPs to separate oil and grease from the effluent.
- COD, pH, coal, and earth particles in the effluent are treated with alum, lime, and polyelectrolyte dosing.
- DSTPs (domestic sewage treatment plants) have been installed to treat domestic effluents, using the activated sludge process.
- TDS and heavy metals are continuously monitored and are well within the permissible limits set by the pollution control board.

Hazardous waste

Burnt oil is generated as hazardous waste from HEMM, and oil emulsion is produced from oil and grease traps of the ETPs. These waste oils are collected, drummed, and sent for disposal through the Metal Scrap Trading Corporation to distillation units possessing necessary license.

Other waste

In coal mines, OB waste is generated in huge quantity. OB dumps are maintained through technical and biological reclamation.

Case study

Assessment of temporal changes in vegetation cover in Singrauli coalfields, based on remote sensing and GIS during 2000 and 2005

Introduction

The NCL is committed to maintaining the ecological balance in the region and has initiated a massive plantation programme in backfilled area, OB dumps, and wasteland. The advent of high resolution, multispectral satellite data has opened up new avenues in the field mapping and monitoring of vegetation cover. Mining has always been associated with the negative impacts it causes on the vegetation of the area. So, for assessing the actual impact of coal mining on the vegetation cover, there is a need for a comprehensive vegetation inventory that includes spatial distribution, aerial extent, location, rate and pattern of change of each vegetation category, and so on. The present study was also undertaken to evaluate the changes in the vegetation cover in Singrauli coalfield in a span of five years, from 2000 to 2005. While assessing the vegetation cover in the coalfield, changes in different categories of land use/cover were also analysed to formulate remedial measures to maintain the ecological balance in the region.

Data source and processing

Raw satellite data was obtained from the NRSA (National Remote Sensing Agency), Hyderabad, which was complemented with secondary (ancillary) and ground data. The processing of the data involved the following major steps.

- Geometric correction, rectification, and geo-referencing
- Image enhancement
- Training set selection
- Signature generation and classification
- Creation/overlay of vector database
- Validation of classified image
- Layerwise theme extraction using GIS (geographic information system)
- Preparation of final vegetation map

The analysis was carried out separately for the entire coalfield spread over 284 km^2 and for the leasehold area of 203 km^2 acquired by the NCL.

Vegetation cover/land use classification in the study area

The land use/vegetation cover classes identified in the Singrauli coalfield region are given in Table 1.

Table 1 Land use/vegetation cover classes in the Singrauli coalfield region

Level I	Level II
Vegetation cover	3.1 Dense forest
	3.2 Open forest
	3.3 Scrub
	3.4 Plantation under social forestry
	3.5 Plantation on overburden dumps
Mining area	5.1 Coal quarry
	5.2 Advance quarry site
	5.3 Barren overburden dump
Agricultural land	2.1 Cropland
	2.2 Fallow land
Wasteland	4.1 Waste upland with/without scrubs
	4.2 Barren rocky land
	4.3 Fly ash pond
Settlements	1.1 Urban
	1.2 Rural
	1.3 Industrial
Waterbodies	6.1 River/streams/reservoirs

Change in the vegetation cover

There was a significant variation in the land use under the vegetation classes, as shown in Table 2.

Dense forest

Forests with a crown density of 40% are included in this class. Dense forest cover over the area decreased due to extensive deforestation by locals, especially in areas where mining had not begun. So, the significant conversion of dense forest to open forest cannot be attributed to mining in the area.

Open forest

Open forest cover also showed a significant decline in the study area, of the order of 4.94 km^2, or 1.74% of the coalfield area. Again this conversion of forests into scrubs and wasteland had taken place in the areas that were not subjected to mining activities.

Scrubs

Scrubs in the coalfield area increased, as the dense and open forests were reported to be converting into scrubs. Even fallow land in some areas had given way to scrubs.

Table 2 Change in vegetation cover

Vegetation cover	2000		2005		Change	
	Area (km²)	% total	Area (km²)	% total	Area (km²)	% total
Dense forest	35.46	12.49	25.81	9.08	(–)9.65	(–)3.39
Open forest	55.60	19.58	50.66	17.83	(–)4.94	(–)1.74
Scrub	26.92	9.48	31.87	11.22	(–)4.95	(–)1.74
Social forestry	35.50	12.50	40.58	14.29	(–)5.08	(–)1.79
Plantation on OB dump	21.75	7.66	28.72	10.11	(–)6.97	(–)2.45
Total	175.23	61.70	177.64	62.54	(–)2.41	(–)0.85

Social forestry

Plantation undertaken on wastelands and along roadsides, and the green belt developed in the colonies come under social forestry. Vegetation cover under social forestry increased by 1.79% of the coalfield area.

Plantation on the OB dump and backfilled area

The NCL has undertaken large-scale plantation on the OB dumps and backfilled areas for maintaining the ecological balance in the area. There was a consistent increase in this vegetation cover during the study period.

The overall increase in the vegetation cover from 2000 to 2005 was to the tune of 2.41 km^2, which is about 0.85% of the coalfield area.

Figure 1 shows the change in the vegetation cover in the Singrauli coalfield area during the study period.

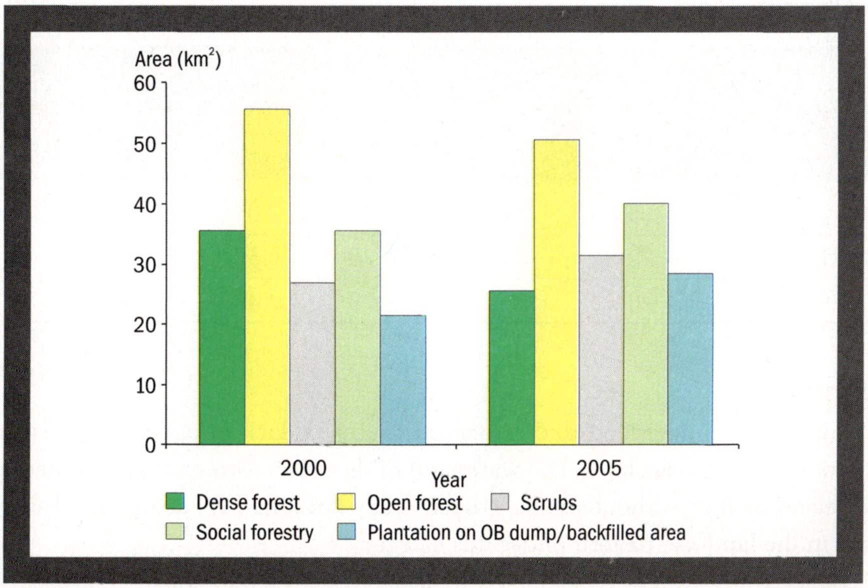

Figure 1 Change in the vegetation cover (from 2000 to 2005)

Mining area

The mining area was categorized into coal quarry, advance quarry site, and barren OB dumps. Table 3 shows the change in the land use pattern under these categories.

The table reveals that the mining area that was 31.42 km^2 in 2000 increased to 36.60 km^2 in 2005, an increase of 1.83% of the coalfield area.

Table 3 Change in the land use pattern in the mining area

	2000		*2005*		*Change*	
Mining area	*Area* (km^2)	*% total*	*Area* (km^2)	*% total*	*Area* (km^2)	*% total*
Coal quarry	22.33	7.85	24.85	8.75	(+)2.52	(+)0.89
Advance quarry site	3.74	1.31	4.79	1.68	(+)1.05	(+)0.37
Barren OB dump	5.35	1.88	6.96	2.35	(+)1.61	(+)0.57
Total	31.42	11.04	36.60	12.77	(+)5.18	(+)1.83

Agricultural land

Land primarily used for farming for growing food, fibre, and other commercial and horticultural crops is termed as agricultural land. It includes both cropland (irrigated and unirrigated) and fallow land (land used for cultivation, but temporarily left unfarmed). Table 4 shows the change in the agriculture land during the study period.

As can be seen from the table, the total agricultural land had decreased by 1.24 km^2, or 0.43% of the coalfield area. This decrease was due to the infrastructure development in the region as well as conversion of some agricultural land into wasteland.

Table 4 Change in the agriculture land

	2000		*2005*		*Change*	
Agricultural land	*Area* (km^2)	*% total*	*Area* (km^2)	*% total*	*Area* (km^2)	*% total*
Cropland	5.71	2.01	8.00	2.81	(+)2.29	(+)0.81
Fallow land	22.51	7.91	18.98	6.67	(–)3.53	(–)1.24
Total	28.22	9.92	26.98	9.48	(–)1.24	(–)0.43

Wasteland

The land that remains degraded due to natural causes or lack of appropriate water and oil management is termed wasteland. The wasteland in the study area can be classified into three types: wasteland with or without scrubs, barren rocky land, and fly ash pond. Table 5 shows the change in the land use pattern under this category.

It is evident from the table that there has been a significant reduction in the wasteland, of the order of 8.57 km^2. This is because some of the wasteland has been converted to vegetated area and some even to mining area.

Table 5 Change in the wasteland

Wasteland	2000		2005		Change	
	Area (km^2)	% total	Area (km^2)	% total	Area (km^2)	% total
Cropland	5.71	2.01	8.00	2.81	(+)2.29	(+)0.81
Wasteland with or without scrubs	33.45	11.76	24.09	8.47	(–)9.36	(–)3.29
Barren rocky land	3.86	1.36	4.56	1.60	(+)0.70	(+)0.25
Fly ash pond	0.21	0.07	0.30	0.10	(+)0.09	(+)0.03
Total	37.52	13.19	28.95	10.18	(–)8.57	(–)3.01

Settlements

Settlements include urban settlement, rural settlement, and industrial settlement. Table 6 gives an account of the change that occurred in this category during the study period.

Table 6 Change in settlements

Settlements	2000		2005		Change	
	Area (km^2)	% total	Area (km^2)	% total	Area (km^2)	% total
Urban settlement	10.09	3.54	11.76	4.13	(+)1.67	(+)0.58
Rural settlement	0.74	0.26	0.57	0.20	(–)0.17	(–)0.06
Industrial settlement	0.23	0.08	0.38	0.13	(+)0.15	(+)0.05
Total	11.06	3.88	12.71	4.46	(+)1.65	(+)0.57

It is clear from the table that the urban settlement within the study area increased during the study period, while the rural settlement decreased, probably due to the migration of the rural population to urban areas. The industrial settlement has also witnessed a slight increase.

Waterbodies

The area under waterbodies in the study area increased by 0.57 km^2 during the study period.

Benefit

The project gives an overview of the change in the vegetation cover and land use pattern over the five years, from 2000 to 2005. Such analyses help in maintaining the ecological balance of the region, mitigating the adverse impacts of mining. Such projects also help in adopting sustainable mining practices.

CASE STUDY AT A GLANCE

- The NCL (Northern Coalfields Ltd), a subsidiary of the CIL (Coal India Ltd), was formed in 1986. Presently, the coal mining activities of the company are spread over an area of 100 km^2.
- The present study was undertaken to evaluate the changes in the vegetation cover in the Singrauli coalfield in a span of five years, from 2000 to 2005.
- While assessing the vegetation cover in the coalfield, changes in different categories of land use/cover were also analysed to formulate remedial measures to maintain the ecological balance in the region.
- Vegetation cover increased during the study period due to social forestry efforts and large-scale plantation undertaken on the OB dumps and backfilled areas.
- Dense forest and open forest decreased due to deforestation by locals.
- Mining area showed an increase in a span of five years, leading to an increase in coal production.
- Large stretch of agricultural land is being converted into mining area.
- Wasteland decreased over the study period, as it was brought under vegetation and mining.
- Settlement area increased due to rapid urbanization and industrialization.
- Area under waterbodies also increased.

CASE STUDY 30 Singareni Collieries Company Ltd

About the company

Andhra Pradesh is endowed with rich coal reserves, mainly in the four districts of Adilabad, Karimnagar, Khammam, and Warangal. The SCCL (Singareni Collieries Company Ltd) is the only company in the entire South India having operations in these four districts, catering to the needs of coal-based industries. It is the second largest public sector undertaking in the state of Andhra Pradesh in terms of employing 71 762-strong workforce.

The company deals with the extraction of coal from underground as well as opencast mines. Presently, the company is operating 37 underground mines and 13 opencast mines and has achieved a record coal production of 40.06 MT (million tonnes) for 2007/08. Having posted a net profit of Rs 1761.6 million during the same period, as against Rs 1170 million in the previous year, the company has fixed a target of 43.56 MT of coal production for 2008/09.

Environmental profile

Departments

The SCCL is an environmentally conscious, responsible, and proactive company. It has taken several measures for controlling pollution and has undertaken extensive plantation activities in the mine area. The company has an environment department in place, which looks after statutory compliance, coordinates with government agencies, and monitors the implementation of environmental management plans and other environment protection measures approved by the MoEF (Ministry of Environment and Forests) or the Andhra Pradesh Pollution Control Board.

The company also has a separate forestry department, which oversees afforestation activities, timber needs, forestland diversion, conservation practices, and so on.

Air pollution

The main air pollutants emitted are SPM (suspended particulate matter), RSPM (respirable suspended particulate matter), SO_2 (sulphur dioxide), NO_x (oxides of nitrogen), and fugitive dust and gases.

Air pollution control measures

- Haul roads, dump yards, and coal screening plants are sprayed with water to keep the road surface wet and prevent the dust from rising due to the movement of dumpers and other vehicles. Five water sprinklers of 28 kilolitres capacity are used for this purpose (Figure 1).
- The entire 9.2-km stretch of the approach road connecting mine, coal handling plant, and workshop has been black topped to prevent dust from being airborne.
- Periodical maintenance of heavy earth moving machinery and other vehicles is carried out to control noxious gases in exhaust emissions.

Figure 1
Mobile water sprinkler of 28 kilolitres capacity

- To reduce air pollution load due to the burning of coal and wood for domestic purposes, free coal supply to employees is replaced with free LPG (liquefied petroleum gas) supply.
- Wet drilling methods are employed.
- Coal transport trucks are covered with tarpaulin sheets for preventing the spillage of coal during transport, which reduces dust pollution.
- Proper blasting pattern with optimum explosive charge and delay action detonators are used to control the emission of noxious gases and dust while blasting.

Noise pollution control measures

- Noise levels are recorded at different locations in and around the project.
- Height of fall is minimized at all coal transfer points, and bins and chutes are lined internally.
- Plantation is undertaken to attenuate noise.
- Operators working in areas with high noise levels are encouraged to use earplugs.
- Frequent maintenance and tuning of machinery are carried out to minimize noise.
- Blasting operations are carried out with proper hookup and optimum explosive charge.
- A green belt comprising species with rich canopy has been developed around the lease area and along the roads, which further attenuates the noise level.

Water pollution

The main water pollutants present in the effluent generated from the facility are oil and grease and total suspended solids.

Water pollution control measures

- The mine discharge water is passed through sedimentation tanks for the removal of suspended solids, after which the treated water is utilized directly for watering the plantation, dust suppression, washing heavy earth moving machinery, and so on.

- Oil and grease are removed by passing the effluent through the effluent treatment plant.
- The chemical characteristics of water discharged from different mines are monitored periodically, and it is ensured that the concentrations of all the parameters are within the prescribed limits, so that the water can be reused.
- Run-off water from the overburden dump is treated in settling tanks to remove suspended solids.
- Garland drains with settling tanks are provided around the quarry and overburden dumps for the natural draining of water during monsoons.
- Check dams/rock fill dams have been constructed at strategic points around the overburden dump to reduce soil erosion.

Hazardous waste

Hazardous waste is generated in the form of used oil, used batteries, scrap iron, worn out bearings, empty oil and grease drums, empty explosive cartons, used tyres, used wire rope, conveyor belt, and so on.

Iron scrap is reused at the unit with some of the scrap sold out occasionally. Old tyres of dumpers, trucks, jeeps, and old batteries are also sold out periodically to the agencies authorized by the Andhra Pradesh Pollution Control Board. Vermicomposting is used in the hospital area to treat the hazardous waste generated by the hospital.

Other waste

In opencast coal mines, the main waste generated is the superincumbent strata above coal seams, termed as overburden. The overburden contains hard sandstone and subsoil. Biological reclamation is the stabilization of overburden dumps by vegetation. The burden is stabilized by spreading topsoil over all dumps and slopes, constructing toe wall around the overburden dumps, constructing check dams/rock fill dams, constructing drains over the dumps and all around the external dumps, and so on.

Case study

Environmental initiatives and eco-friendly mining at Ramagundam Opencast-III Project

Ramagundam Opencast-III Project is located in the central part of Ramagundam coal belt of Godavari Valley coalfields (Figure 2). The mine started producing coal in 1990, with a rated capacity of 2.75 MTPA (million tonnes per annum). The project will be operational for 21 years, with an annual operating capacity of 4.30 MTPA and peak production capacity of 5.0 MTPA. The project was accorded environmental clearance by the MoEF in 2008 for further expansion.

Eco-friendly mining

The overburden in the project is dumped in the demarcated location. As the dump height increases, soil erosion starts taking place, particularly from the slopes of the dump. Since

Figure 2
Ramagundam Opencast-III Project

the dump is still under formation, stabilization by vegetation cannot be taken up. Emphasis is placed on restoring soil organic matter, nutrients, and vegetation cover to accelerate the natural recovery process.

Ecological restoration

The goal of ecological restoration is to develop a long-term sustainable ecosystem native to the area where mining is usually carried out. It aims at restoring the degraded system, so that it is productive, aesthetically pleasing, and valuable.

Resource conservation

The major portion of the overburden dump is relocated to de-coaled area. The land required for the extension project was minimized by planning the dumping of overburden in existing dumps. No forestland was diverted for this project.

Slope stability analysis

Topsoil management

About 7.64 million m^3 of topsoil was removed as on 31 March 2008, of which 1.19 million m^3 of soil was spread on external dumps, 0.3 million m^3 of topsoil was used in backfilled area reclamation, 2 million m^3 of topsoil was used for the management of subsidence areas of adjoining mines, and 3.84 million m^3 of topsoil has been preserved for future reclamation.

Biological engineering

Engineering principles

Concerted efforts were taken to make the overburden stable by constructing various engineering structures such as toe wall around the external dump, settling tanks, check dams, rockfill dams, gabions, cribs, drains on the top of the dump, contour trenches over dump slopes, and so on. Apart from this, gullies and channels were constructed for the smooth drainage of rainwater, minimizing erosion.

Biological principles

The aim of biological measures is to stop soil erosion by forming vegetative barriers and increasing groundwater level. These measures help in the reclamation process and in restoring the natural faunal and floral elements. Some of the major species used for overburden reclamation include *Albegia odorati*, *Pongamia pinnata*, *Termilania arjuna*, *Acacia nelotica*, *Azadirechta indica*, *Ficus religiosa*, and so on. Figure 3 gives a view of the reclaimed project site.

Figure 3
A view of the reclaimed project site

Pollution control measures

Dust pollution control

- Haul roads, dump yards, and coal screening plant are sprayed with water to prevent the dust from rising.
- The entire length of the approach road has been black topped.
- Wet drilling is undertaken.

Gaseous pollution control

- Maintenance works of heavy earth moving machinery and other vehicles are undertaken regularly to control noxious gases.
- Proper blasting pattern with optimum explosive charge and delay action detonators are used to control the emission of gases while blasting.

Water pollution control measures

- Workshop effluent is treated in 'oil and grease trap' before releasing it (Figure 4).
- Mine water is treated in settling ponds inside the quarry.
- Waste water generated in office complexes is treated in septic tanks and used for watering the plantation developed in and around the mine area.
- Rainwater harvesting is practised to conserve water.
- Rainwater is conserved in a lake spread over an area of 3.5 hectares, with a capacity of 40.5 million gallons. The lake also collects run-off and storm water from the mine.

Figure 4
Oil and grease trap

Noise pollution control measures

- To dampen the noise levels at coal handling plants, impact rollers are provided at transfer points.
- Height of the fall is minimized at all coal transfer points, and bins and chutes are provided with internal lining.
- A green belt has been developed around the mine area to attenuate the noise level.
- Regular maintenance and tuning of machinery are done.
- Blasting operations are carried out using NONEL delay detonating system to control the ground vibrations.

Soil erosion control measures

- Rock toe wall is constructed to prevent silt flow in the run-off water from the overburden dump (Figure 5). The dump is also covered with vegetation (Figure 6).
- Rockfill dams/check dams have been constructed in the garland canals.
- Settling ponds for silt have been constructed.

Figure 5
Toe wall around the overburden dump to control soil erosion

Figure 6
One of the overburden dumps with toe wall, garland canal, and green cover

- Two RCC slope drains have been constructed at narrow gullies on the dumps.
- Two slope drains have been constructed, with pitching in wide gullies on the dumps.

Cost–benefit analysis

An amount of Rs 92.69 million has been spent on implementing various environment-friendly practices and pollution control measures. These have resulted in the increase in coal production, as the mine produced 4.56 MT and 4.06 MT of coal in 2006/07 and 2007/08, respectively, which is 8.1 MT more than the rated capacity of the mine.

Environmental benefits

- Conservation of natural resources
  - *Coal* Five underground mines have been converted to opencast, considering adverse geo-mining conditions and safety issues.
  - *Land* The land required for dumping of overburden is minimized by managing the existing dumps.
  - *Timber* The company is raising Eucalyptus plantation to meet the timber needs in underground mines.
  - *Water* Water is conserved in waterbodies like lakes, ponds, and so on. Fish are harvested in such waterbodies.
- The overburden dumps are reclaimed, which results in the restoration of the ecological balance.
- The emission of carbon dioxide, a greenhouse gas, has been reduced to a great extent.
- Various environment-friendly practices and pollution control measures implemented by the facility have resulted in a significant increase in the coal production.

CASE STUDY AT A GLANCE

- The SCCL (Singareni Collieries Company Ltd) is the only coal mining company in South India catering to the needs of coal-based industries.
- The company deals with the extraction of coal from underground as well as opencast mines.
- An amount of Rs 92.69 million has been spent on implementing various environment-friendly practices and pollution control measures.
- These practices have led to conservation of natural resources like coal, land, timber, and water.
- The emission of carbon dioxide, a greenhouse gas, has been reduced to a great extent.
- Various environment-friendly practices and pollution control measures implemented by the facility have resulted in a significant increase in the coal production.

CASE STUDY 31 NHPC Ltd

About the company

NHPC Ltd, formerly known as National Hydroelectric Power Corporation Ltd, is a Government of India enterprise, which was incorporated in 1975 with an authorized capital of Rs 2000 million, with an objective to plan, promote, and organize an integrated and efficient development of hydroelectric power. Later, the NHPC expanded its objectives to include other sources of energy like geothermal, tidal, wind, and so on. Since its inception in 1975, the NHPC has grown to become one of the largest organizations in the field of hydropower development in the country.

During 2007/08, the NHPC had a sales turnover of Rs 23 010 million, with a net profit of Rs 10 040.9 million.

The Uri Hydroelectric Project is located in Jammu and Kashmir. The project is a run-of-the-river scheme on the River Jhelum, located on the downstream of lower Jhelum Power Station. The project comprises a barrage (Figure 1), cut and cover, desilting basin, open channel, 10.65-km-long head race tunnel, two pressure shafts, an underground powerhouse, and 2.06-km-long tail race tunnel, with outfall at the Bandi village, near Uri town. The project has an installed capacity of 480 MW (4 × 120 MW) under Stage I, with a provision to install an identical capacity under Stage II when the additional regulated water becomes available with the construction of Wular Barrage and Kishenganga Hydroelectric Project. The project is designed for the annual generation of 2663 million units in 90% dependable year.

Environmental profile

The NHPC's objective 'to execute and operate project in a cost-effective, environment-friendly, and socio-economically responsible manner' reflects its concern towards the environment.

Figure 1
A view of Barrage upstream

It has an Integrated Management System Policy that aims at the sustainable development of hydropower and water resources, with special focus on the environment. As per its Environment Policy, the Uri Power Station aims at addressing environmental concerns in hydropower development, commits to comply with legal environmental procedures and prevent pollution from all the activities, and endeavours to optimally utilize the natural resources.

The company has a separate environmental cell at both corporate and facility levels, which works closely with the Ministry of Environment and Forests and state pollution control board. It undertakes EIA (environmental impact assessment) and SIA (social impact assessment) studies, along with remote sensing and GIS (geographic information system) studies. It is also responsible for the monitoring of environmental parameters.

The company has an IMS (International Management System), which adheres to the international standards like ISO 9001: 2000, ISO 14001: 2004, and OHSAS 18001: 2007 as per PAS 99: 2006.

As part of the Environmental Management System, various environmental management programmes aiming at energy and paper conservation have been implemented at the Uri Power Station.

The Uri Power Station uses water for generating power. So, there is no emission of polluting gases into the atmosphere. There is no water pollution as well. Elaborate operational control procedures are followed for collecting and disposing of the waste oil, a hazardous waste, from various units to authorized recyclers. Thus, care is taken to ensure that there is no spillage or leakage of oil. No other industrial waste is generated from the facility.

Case study

Implementation of environmental management systems and the post-construction environmental and social impact assessment studies of the Uri Power Station, J&K: a stride towards sustainable development

Introduction

Environment conservation is a cross-cutting issue in the overarching framework of development, wherein sustainability of the natural resources is of prime concern. The Uri Power Project is a project of considerable magnitude and involved the construction of huge civil and electrical structures (Figure 2). The construction of a project of such a huge scale inadvertently causes adverse environmental impacts during the construction and operation phases. The Uri Power Project has taken many corrective measures to reverse the negative impacts it had caused on the environment. Environmental management plans with respect to CAT (catchment area treatment), compensatory afforestation, restoration of muck disposal sites, landscaping, and movement of migratory species of fish were implemented. A post-construction impact assessment, SIA, was also undertaken to assess and evaluate the Human Development Index, quality of life, and need and aspirations of the project-affected persons and locals.

Figure 2
Construction of Uri Barrage

Catchment area treatment

Large-scale deforestation, overexploitation of forest resources, encroachment, and overgrazing on the land, coupled with weak geological formations and rugged topography, had made the catchment area vulnerable to soil erosion. This led to the formation of deep gullies, landslides, loss of fertile top soil cover, and sedimentation of the waterbody. To arrest soil erosion in the catchment, 13 180 hectares of the catchment area, covering nine micro-watersheds, was brought under the CAT plan. The plan included intensive soil conservation measures to keep the soil cover in the area intact and intensive land development strategies, along with torrent control and protection measures. Under these measures, bench terracing, contour bunding, construction of check dams and gabions, and large-scale plantation were undertaken.

Restoration plan

The project area was subjected to extensive excavation during the construction phase, which had affected the topography of the area. Besides, approximately 5.5 million m^3 of muck was generated during the process, which was dumped at 10 designated sites (Figure 3). To reclaim and restore these dumping sites, a restoration plan was implemented on 95 hectares of muck dumping sites by the NHPC, in association with NEERI (National Environment Engineering and Research Institute), under which an integrated biotechnological approach was adopted (Figure 4). The approach comprised following steps.

- Evaluation of spoil tips for their physical and chemical properties to assess their supportive and nutritive capacity.
- Delineation of appropriate blends of organic waste and soil to develop rhizosphere of good supportive capacity.
- Isolation and screening of specialized strains of mycorrhizal fungi, rhizobium, azotobacter, and phosphate solubilizers (biofertilizers inoculum) in accordance with the suitability for the spoil type.
- Mass culture of plant-specific biofertilizers and mycorrhizal fungi.

Figure 3
Muck dumping site

Figure 4
Dumping site at TRT (afforestation work undertaken by using integrated biotechnology)

- Plantation of spoil tip areas (including plain and slope areas), using identified blend and biofertilizer inoculum.

The waste material dumped at spoil tips mainly comprised loose rock fragments, which had been mechanically compacted and properly levelled with suitable slopes (Figure 5). Plant species of high ecological and economic values, which could easily adapt to local environmental conditions, were planted on the soil tip areas at a rate of 1800–2200 saplings per hectare. The species planted included *Populus* sp., *Salix alba*, *Robinia pseudoacacia*, *Ailanthus excelsa*, *Cupressus* sp., and so on. Because of this plantation drive, these dumping sites are now marked by luxuriant tree cover.

Landscape plan

Under the landscape plan, parks and gardens were developed in the project area to enhance its aesthetic appeal, which attract people and children in droves.

Figure 5
Slope stabilization at dumping site of Barrage complex (plantation work started)

Compensatory afforestation programme

The project had used 54.6 hectares of forest land, which had resulted in the felling of about 4000 trees. Compensatory afforestation programme was implemented on 62.7 hectares of degraded forest. Under this programme, more than 100 000 indigenous, fast-growing trees were planted, which could also meet the fodder and fuelwood needs of the locals. The afforestation programme also helped in the natural regeneration of the area. Dense and fruit-bearing trees have also been planted along the approach roads and in project colonies.

Fish ladder

To facilitate the smooth movement of fish, a fishway has been constructed across the Uri Barrage (Figure 6). The fish ladder is a modified version of the weir-type of ladder, referred to as 'pool type with orifices and notched overflow'. This fish ladder was constructed after a detailed survey was carried out to assess the impact of the project on the fish and other fauna of the River Jhelum. The fish species of the river comprise many migratory species like snow trout, which migrate upstream for spawning. The fish ladder ensures that the movement of the fish is unhindered.

SIA

The Uri Power Project has resulted in the overall development of the region and the natives, as revealed by the SIA conducted post-construction.

- As per the baseline socio-economic survey of the project-affected families, conducted at the time of the construction of the project, most of the people were illiterate. But this has changed with the coming up of a school in the project region, which is imparting education not only to children but also to grown-ups (Figure 7).
- There has been a significant development in road and infrastructure facilities in the region.
- Better medical facilities are available in the project area.

Figure 6
Fish ladder

- All the project-affected villages have electricity, and the supply is not erratic anymore.
- Civic amenities like banks and post offices have witnessed a marked improvement in the area.
- The project brought with it employment opportunities for locals and project-affected people, both directly and indirectly.
- There has been a marked rise in the living standards of the people.

Figure 7
Kendriya Vidyalaya at Project Colony

CASE STUDY AT A GLANCE

- NHPC Ltd, formerly known as National Hydroelectric Power Corporation Ltd, is a Government of India enterprise, which was incorporated in 1975 with an authorized capital of Rs 2000 million, with an objective to plan, promote, and organize an integrated and efficient development of hydroelectric power.
- The Uri Hydroelectric Project is located in Jammu and Kashmir. The project is a run-of-the-river scheme on the River Jhelum, located downstream of lower Jhelum Power Station.
- The project is designed for the annual generation of 2663 million units in 90% dependable year.
- To reclaim and restore the dumping sites, where muck generated during the excavation was dumped, a restoration plan was implemented, in association with NEERI (National Environment Engineering and Research Institute), under which an integrated biotechnological approach was adopted. These sites now have luxuriant green cover.
- Under the landscape plan, parks and gardens were developed in the project area to enhance its aesthetic appeal.
- The project had used 54.6 hectares of forest land, which had resulted in the felling of about 4000 trees. Compensatory afforestation programme was implemented on 62.7 hectares of degraded forest, under which more than 100 000 indigenous, fast-growing tree species were planted.
- To facilitate the smooth movement of migratory fish, a fishway has been constructed across the Uri Barrage.
- The Uri Power Project has resulted in the overall development of the region and the natives, as revealed by the SIA (social impact assessment) conducted post-construction.

CASE STUDY 32

Wipro Ltd

About the company

Wipro Infotech is the leading strategic IT partner for companies across India, the Middle East, and Asia-Pacific, offering integrated IT solutions. It plans, deploys, sustains, and maintains the IT lifecycle of people through total outsourcing and providing consulting services, business solutions, and professional services. It is a global service provider, delivering technology-driven business solutions that meet the strategic objectives of its clients. Wipro Infotech is a part of $5 billion Wipro Ltd (NYSE:WIT), with a market capitalization of $24 billion. It has 55+ 'centres of excellence' that create solutions around specific needs of industries. The company delivers unmatched business value to customers through a combination of process excellence, quality frameworks, and service delivery innovation. It is the world's first CMMi Level 5 certified software services company and the first outside USA to receive the IEEE Software Process Award. It is the world's first IT Services Company to use Six Sigma.

Its IT services revenue in 2008 was $1126 million, ahead of its guidance of $1121 million. IT products business recorded a 25% year on year growth in revenues and 45% year on year growth in profit before interest and tax.

Environmental profile

Wipro recognizes that 'the policies that we adopt today will shape not only our lives but also those of future generations.' Therefore, in an effort to address social, economic, and environmental concerns, the company has adopted a comprehensive Environment Policy. Wipro is ISO 14001:2004 compliant.

Environmental risk management and disaster management

As part of the EMS (Environment Management System) and ISMS-27001 Implementation, the company carries out environmental review once in a month. As per the OHSAS, the company has identified various risks involved in the hazardous materials through HIRA (Hazard Identification and Risk Assessment), and as part of ISMS 27001 implementation, the company has a disaster management plan in place.

Environmental programmes

Wipro is an active participant in the Carbon Disclosure Project, or CDP 6. It has also entered into a partnership with WWF (Worldwide Fund for Nature) India to explore the use of information and communication technology to drive sustainable development. The company is also a member of the UN Global Compact.

Resource conservation

As part of the EMS (ISO 14001), the company has taken various initiatives towards energy conservation (5% decrease on year on year), water conservation (10% decrease on year on year), and waste management.

Environmental initiative

Wipro's initiative towards ecological sustainability is called EcoEye. It is a concerted corporationwide initiative to transform the way the company does business and engages all its stakeholders. It signifies the company's endeavour to view every dimension of its business with an 'eco' eye. It is also a growth strategy to meet customers' needs for ecologically sustainable products and services.

Air pollution and its control measures

Diesel generator sets are the source of air pollution. The generators consume about 376 726 litres of diesel per annum, or 1028 litres per day. The stacks are monitored every quarter. Most of the air pollutants like suspended particulate matter, and oxides of nitrogen and sulphur are well within the prescribed limits.

Water pollution and its control measures

The total quantity of raw water used in the company is 69 265 kl (kilolitres) per annum. For flushing, 26 591 kl of recycled water per annum is used. About 32 461 kl of recycled water per annum is used for gardening. So, the total quantity of water used is 128 317 kl per annum. The effluent generated is treated in an STP (sewage treatment plant), as per the directions of the KSPCB (Karnataka State Pollution Control Board). The parameters like BOD (biochemical oxygen demand), COD (chemical oxygen demand), pH, oil and grease, and total suspended solids are monitored and tested once in a month. Monitoring of pH and mixed liquor suspended solids is done every day. The sludge generated is sent for analysis to an approved testing laboratory to determine the presence of heavy metals. The treated effluent thus obtained is used for horticultural and flushing purposes.

Solid waste

Table 1 gives the particulars of the different types of solid waste generated, along with their disposal methods.

Table 1 Types of solid waste generated and their disposal methods

Type of solid waste	Quantity generated per day	Mode of disposal
Paper waste	3 kg	Shredded and sent for recycling
Plastic waste	7.5 kg	Sent to Mythri Seva Samithi
Packaging waste	Nil	Nil
Sludge from the sewage treatment plant	4 kg	Used as manure in gardening
Food waste	109 kg	Used as compost
Vegetable waste	258 kg	Used as compost

Hazardous waste

The hazardous waste generated is in the form of used oil/spent oil, waste oil, oil-soaked cotton, generators' air and oil filters, and e-waste like computers. Used oil/spent oil and waste oil are sent to the CPCB (Central Pollution Control Board)-authorized vendor for recycling. Oil-soaked cotton and filters are sent to the CPCB-authorized vendor for incineration. Finally, e-waste is also sent to the CPCB-authorized vendor for recycling. Wipro launched the GreenWare range of Desktops and Notebooks on 14 June 2007. This range is fully compliant as per RoHS (Restriction of Hazardous Substances) directive.

Noise pollution

Diesel generator sets are the only source of noise pollution. As per the environmental audit, it has been observed that the noise generated from generator sets is within the limits prescribed by the pollution control board. The generator sets are enclosed with acoustics.

Case study: Paper recycling plant

A paper recycling plant at Wipro Ltd has been set up to ensure that no waste paper leaves the facility with printed matter. The plant is in operation since April 2007. The aim is to convert the waste paper into compressed solid blocks, which can be supplied to a paper manufacturer to make stationary items like pads, which, in turn, are supplied back to the company (Figure 1). The plant was set up at a cost of Rs 610 000 and the operational cost per month is Rs 9500. The company has processed 14 tonnes of paper so far, which is equivalent to saving 39.2 tonnes of dry timber and approximately 1 135 623 litres of water.

Figure 1
Eco-friendly pad made out of recycled paper

Aim of setting up the plant

The plant was set up with the following aims.

- To handle the paper waste efficiently and avoid document/data security issues.
- To recycle resources within the campus.

Plant specifications

The capacity of the paper recycling plant is 200 kg/day. It has been established in an area of 400 square feet. Currently, on an average, it recycles about 4680 kg of paper per month.

Figure 2
Paper recycling plant at Wipro Ltd premises

It comprises a paper crusher motor (10 HP/3 Ph), a recycle water pump (3 HP/3 Ph), a paper slurry shifting pump (5 HP/3 Ph), and a hydraulic press pump (1 HP/3 Ph) (Figures 2 and 3). The plant consumes 12.5 units of power per day and 600 litres of water per day.

Figure 3
Model of the paper recycling plant

Process

The hydraulic pulper is fed with waste paper and water. It crushes the paper into slurry/ pulp. This pulp is pumped into the hydraulic basket press, which converts the pulp into cakes of 25 kg each. These cakes are given to large paper manufacturers for converting them into pads.

Monitoring

One operator monitors the plant 8 hours per day for a week or so, depending on the load.

Cost–benefit analysis

Paper generated at the Electronic City campus	: 300 kg
Expected waste paper generation from other BDC locations	: 150 kg
Capacity of the plant per day	: 200 kg
Average pulp generated per day (wet cakes)	: 420 kg
Cost of pulp per kg	: Rs 2.25
Revenue per day	: Rs 945
Savings per day	: Rs 529
Savings per month (25 days)	: Rs 13 754
Savings per annum	: Rs 165 048

Payback period

Payback period = Net cost of the plant/net saving per month
= 610 000/13 754
= 44 months, or 3.7 years

Benefits

- *Conservation of resources* The company has processed 14 tonnes of paper so far, which is equivalent to saving 39.2 tonnes of dry timber and approximately 1 135 623 litres of water.
- *Savings* Savings per annum due to the installation of the paper recycling plant are to the tune of Rs 165 048

Benefits of recycling

- *Conservation of natural resources* Paper recycling conserves natural resources like forests and water. Recycled paper uses 55% less water.
- *Emission reduction* The recycling process prevents greenhouse gas emissions.
- *Water pollution control* Recycled paper reduces water pollution by 35%, air pollution by 74%, and eliminates toxic wastes.
- *Energy conservation* Paper recycling saves energy, which is to the tune of 60%–70% savings over virgin paper manufacture.
- *Raw materials* The process of recycling supplies valuable raw materials to industry.
- *Waste disposal* Paper recycling reduces the need for new landfills and incinerators.

Case study

Implementation of a rainwater harvesting system at Wipro Ltd, Electronic City

Water is a scarce commodity in the Electronic City, where Wipro Ltd is located, which is dependent on groundwater resources to meet its water demands. This overuse of groundwater has led to its depletion, forcing the city to look for alternative sources of water.

To reduce its dependence on groundwater and utilize the rainwater going waste into drains, the company decided to implement a rainwater harvesting system in its facility. The feasibility study of the rainwater harvesting system was carried out with the help of an external consultant, who proposed the setting up of a storage facility of 350 kl/day.

Data on the average rainfall in the area for the past three years was collected. This data, along with the total open area and rooftop area of the towers, was analysed to arrive at the volume of rainwater that could be harvested.

Four underground collection sumps of 350 kl capacity were constructed at strategic points. Relatively pure rooftop water was made to flow to two sumps through PVC (polyvinyl chloride) pipes. Storm water drains drained the surface water to other two sumps.

To handle the excess rainwater during heavy downpour, which would otherwise overflow from the storage tanks into the facility, three groundwater recharging pits (6 feet wide and 10 feet deep) were constructed near the underground collection sumps. The excess water collects in these pits, recharging the groundwater. Rain gauges are used to measure and monitor the rainfall. About 6900 kl of rainwater is being harvested annually.

The total rainwater harvested is calculated using the following formula.

Rainwater harvested (rooftop) + rainwater harvested (open area) = run-off coefficient (0.9) × rooftop area (in m^2) + run-off coefficient (0.45) × rainfall (in m) × open area (in m^2)
Actual collection of rainwater in the sumps is measured using a dipstick, and the variation is found to be of the order of 15%.

The harvested water is used in cooling towers, in cafeteria, and for landscaping.

Cost–benefit analysis

Quantity of rainwater harvested annually	: 6900 kl
Cost of raw water per kl	: Rs 30
Cost of the project	: Rs 270 000
Annual savings (rainwater harvesting)	: Rs 207 000
Payback period	: 1.3 years

Benefits

- *Conservation of water* Annual water conservation to the tune of 6900 kl, equivalent to 15 days of water consumption of the facility, has been achieved.
- *Groundwater recharging* Due to the percolation of excess rainwater into the ground through recharging pits, the groundwater level in the area has improved.
- *Saving in power consumption* About 2600 units of electricity is being saved annually, which were earlier being used to pump groundwater.
- *Savings* Annual savings due to rainwater harvesting are to the tune of Rs 207 000.

CASE STUDY AT A GLANCE

- Wipro Infotech is the leading strategic IT partner for companies across India, the Middle East, and Asia-Pacific, offering integrated IT solutions.
- Its IT services revenue in 2008 was $1126 million, ahead of its guidance of $1121 million.
- A paper recycling plant was set up at Wipro facility in 2007 at a cost of Rs 610 000.
  - The company has processed 14 tonnes of paper so far, which is equivalent to saving 39.2 tonnes of dry timber and approximately 1 135 623 litres of water
  - Annual savings realized per annum are to the tune of Rs 146 250.
- To reduce its dependence on groundwater and utilize the rainwater going waste into drains, the company decided to implement rainwater harvesting system in its facility.
- Four underground collection sumps of 350 kl capacity were constructed at the strategic points.
- To handle the excess rainwater during heavy downpour, which would otherwise overflow from the storage tanks into the facility, three groundwater recharging pits were constructed near the underground collection sumps.
  - Annual water conservation to the tune of 6900 kl, equivalent to 15 days of water consumption of the facility, has been achieved.
  - Due to the percolation of excess rainwater into the ground through recharging pits, the groundwater level in the area has improved.
  - About 2600 units of electricity are being saved annually, which were earlier being used to pump groundwater.
  - Annual savings due to rainwater harvesting are to the tune of Rs 207 000.

CASE STUDY 33 Honeywell Electrical Devices and Systems India Ltd

Figure 1
Honeywell Electrical Devices and Systems India Ltd, Chennai

About the company

Honeywell International is a Fortune 100 global diversified technology and manufacturing leader (Figure 1). Each of the company's four businesses – aerospace, automation and control solutions, transportation systems, and specialty materials – has operations in India. Honeywell has set up state-of-the-art manufacturing and engineering operations for its automation, turbocharger, and refining businesses, and operates its global centres of excellence for research, product development, and innovation in India. Honeywell directly employs more than 10 000 people based in Delhi, Pune, Bangalore, Hyderabad, Chennai, Gurgaon, Madurai, and several other cities.

Honeywell Electrical Devices and Systems, or ED&S, based in Chennai, is part of the Honeywell International Group and was the first company in India to introduce modular electric switches through its 'MK' brand name. It manufactures electrical wiring devices and accessories. In 2007, the company commissioned its second manufacturing facility in the country at Hope Town near Dehradun.

Annual turnover of the company in 2006/07 was Rs 840 million.

Environmental profile

The company has a joint Health, Safety and Environment Policy, under which it is committed to ensure a clean, green, and safe working environment. It optimizes resources like water, raw materials, and electricity. It monitors and effectively manages air emissions, effluents, and solid waste while complying with all relevant legislative norms.

The facility does not emit any air pollutant. Diesel generator sets are used during power failure. The emissions from these are monitored periodically and are generally within standards.

The main water pollutants present in the waste water generated from the facility are BOD (biochemical oxygen demand), pH, oil and grease, total dissolved solids, and total suspended solids. The waste water is treated in the sewage treatment plant, and the treated water is reused in the facility. Oil and grease are reclaimed from the waste water.

The facility also generates spent oil and batteries, which are sold to authorized recyclers. Plastic waste is recycled in-house.

Case study

Eco-friendly practices in the manufacture of electrical and electronic products

In this technology-driven world, the demand for electrical and electronic products is increasing by the day. These products have a very short life cycle, as there is constant innovation and upgradation in this segment. Also, there is a growing awareness on the adverse impacts these products have on the environment during their manufacture, life cycle or when they become obsolete. Thus, the manufacturing companies are increasingly realizing that there is a need for introducing eco-friendly technologies in their manufacturing processes and are also ensuring that their products are environment friendly. The main environmental issues that need to be addressed in the manufacturing and disposal of electrical and electronic waste include the following.

- Improper disposal of end-of-life electrical and electronic products.
- Land pollution due to the disposal of lead-based products.
- Air pollution because of the incineration of e-waste.
- Water pollution due to the disposal of process solution into waterbodies.
- Adverse impacts of hazardous waste on water, air, and land.

Honeywell has implemented some innovative measures with a view to restrict the use of hazardous substances and make its operations environment friendly.

- The facility is using lead-free soldering for PCB (printed circuit board) and electronic assembly.
- The usage of mercury in thermostats, sensors, and relays in switches and discharge lamps has been discontinued.
- The company avoids the use of cadmium-based colorants.
- The company has replaced hexavalent chromium with trivalent chromium in the plating process.
- The use of flame retardants like polybrominated biphenyls and polybrominated diphenyl ether in PCBs, connectors, and plastic covers has been discontinued.

- To mitigate dioxin emissions during the burning of PVC (polyvinyl chloride) blisters in the conventional heat sealable RPVC blister packaging, innovative cold sealing technology with PET (poly(ethylene terephthalate)) blister has been introduced.
- Hazardous waste generated from the manufacturing unit is sent to only the authorized recyclers.
- The company has implemented a product risk assessment process during product development, which covers both health and environment aspects. For this, a separate SOP (standard operating procedure) has been developed, which ensures that the products manufactured comply with laws, regulations, and government standards that relate to the design, manufacture, distribution, use, and disposal of these products.

Benefits

- There is a reduction in the generation of hazardous waste during the manufacturing process and disposal of the product.
- Because of the reduction in the generation of waste, there is less pressure on landfills.
- As there is a reduced usage of hazardous carcinogenic materials, there are no adverse impacts of the facility operations on the health of employees and communities handling the products.
- There has been a decrease in the pollutants like dioxins that are being emitted into the air.
- The company has integrated the concept of recycling in its activities, thus conserving raw material.
- Ecobliss packaging used by the company for its products has set an eco-friendly trend (Figure 2).
- The company insists that its suppliers also adopt eco-friendly measures, thus extending its green practices to others involved in the value chain.

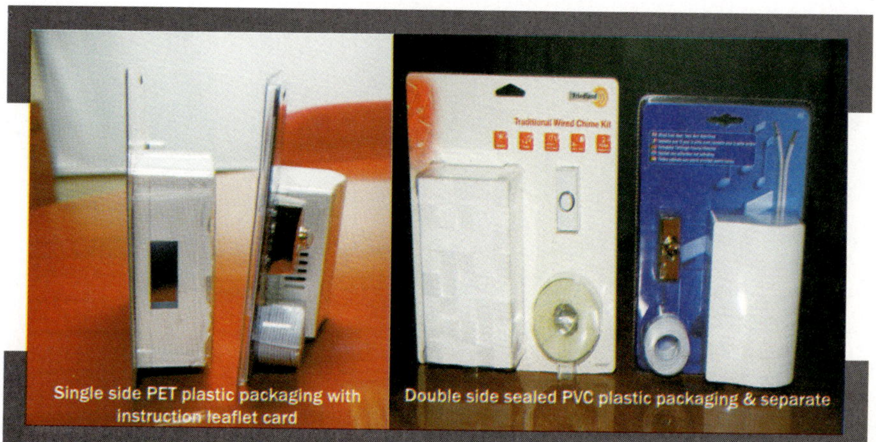

Figure 2 Ecobliss packaging using cold sealing process and single side PET poly(ethylene terephthalate) as against the conventional two side PVC (polyvinyl chloride) blister packaging

CASE STUDY AT A GLANCE

- Honeywell Electrical Devices and Systems, or ED&S, based in Chennai, was the first company in India to introduce modular electric switches through its 'MK' brand name.
- It manufactures electrical wiring devices and accessories.
- Honeywell has implemented some innovative measures with a view to restrict the use of hazardous substances and make its operations environment friendly.
- There is a reduction in the generation of hazardous waste during the manufacturing process and disposal of the product.
- Because of the reduction in the generation of waste, there is less pressure on landfills.
- As there is a reduced usage of hazardous carcinogenic materials, there are no adverse impacts of the facility operations on the health of employees and communities handling the products.
- There has been a decrease in the pollutants like dioxins that were earlier being emitted into the air.
- The company has integrated the concept of recycling in its activities, thus conserving raw material.
- Ecobliss packaging used by the company for its products has set an eco-friendly trend.
- The company insists that its suppliers also adopt eco-friendly measures, thus extending its green practices to others involved in the value chain.

CASE STUDY 34 Indian Oil Company Ltd

About the company

The Indian Oil Company Ltd, established in 1959, merged with the Indian Refineries Ltd in 1964 to form the Indian Oil Corporation Ltd, or IOCL. The IOCL has four divisions: refineries, marketing, pipelines, and R&D (research and development). The company controls 10 of India's 19 operating refineries (including subsidiary unit refineries) and has a combined rated capacity of 60.20 MTPA (million tonnes per annum). The refineries also include two refineries of subsidiary Chennai Petroleum Corporation Ltd.

The Corporation's cross-country network of crude oil and product pipelines, spanning about 9300 km, is the largest in the country, meeting the vital energy needs of the consumers in an efficient, economical, and environment-friendly manner.

IndianOil's world-class R&D Centre is perhaps Asia's finest. Besides being a pioneer in lubricant formulation, refinery processes, pipeline transportation, and alternative fuels, the Centre is also the nodal agency of the Indian hydrocarbon sector for ushering in hydrogen fuel economy in the country. It is in the process of setting up a commercial hydrogen–CNG (compressed natural gas) station at an IndianOil retail outlet in New Delhi this year. It commissioned a biogas plant and a biomass gasifier plant during 2007/08 for conducting research on energy-efficient biogas.

The major products of the company include LPG (liquefied petroleum gas), motor spirit, naphtha, aviation turbine fuel, kerosene, diesel, bitumen, and so on. The Corporation sold 59.29 MT (million tonnes) of petroleum products during 2007/08, as compared to 54.84 MT during the previous year. These include natural gas, the sale of which has gone up to 1.74 MT in 2007/08 from 1.48 MT in the previous year. In addition, product exports in 2007/08 rose to 3.33 MT from 3.13 MT in the previous year.

IndianOil's gross turnover for 2007/08 reached a new high of Rs 2 474.79 billion, up by 12.1% from the previous year.

Environmental profile

IndianOil's vision 'a major diversified, transnational, integrated energy company, with national leadership and a strong environment conscience, playing a national role in oil security and public distribution' kind of sums up its commitment towards preserving the environment. The company has a combined Safety, Health, and Environment Policy that lays emphasis on protecting the environment and ensuring sustainable development. The Environment Policy at the facility level is part of ISO 14001 certification. The environment management systems of all IndianOil refineries, pipeline installations, and major marketing installations/terminals are certified to ISO 14001 standards.

IndianOil refineries have adopted various measures for the control of gaseous emissions. These include use of low-sulphur fuel oil, desulphurization of refinery fuel gas, tall stacks for better dispersion of flue gases, advanced process control systems, and energy conservation measures to reduce fuel consumption.

All refineries have been provided with efficient effluent treatment plants consisting of physical, chemical, biological, and tertiary treatment facilities. 'Oilivorous-S' and 'Oilivorous-A' technologies are being used for the treatment of oily sludge and acid tar, respectively. The treated effluent is far superior to the stipulated MINAS (Minimum National Standards) and quantum standards. The company recycles and reuses the treated effluent for various purposes like firewater make up, cooling water make up, cutting coke in delayed coking units, and for watering the green belts and eco-parks set up in the premises of refineries. This has led to reduced discharge of treated effluent and freshwater consumption.

Oily sludge is the hazardous waste generated from the refineries. It is processed in sludge melting pits, where petroleum is recovered. The recovered oil is reprocessed, while the residual sludge is disposed of through biodegradation and natural weathering.

Case study

Development of alternative non-fossil fuels for reducing vehicular emissions

Bio-diesel – an introduction

Vehicles have been acknowledged as one of the major causes of air pollution. Increasing awareness on vehicular pollution and stringent emission norms have necessitated the use of better quality fuel. Refiners have been looking at various alternatives like bio-desulphurization of diesel or using CNG to tackle sulphur oxide emissions. Indian refineries have spent millions in upgrading their refining facilities to produce low-sulphur (<500 PPM) diesel. CNG has already been introduced in major metros.

An alternative green fuel is bio-diesel, which is made from the trans-esterification of vegetable oil with alcohol. It is increasingly being used as a blend with conventional diesel. In the USA and Europe, soybean oil, sunflower oil, and rapeseed oil are used as feedstock for the production of bio-diesel. India has the potential of producing bio-diesel from non-edible oil plants like *Jatropha curcas*, *Pongamia pinnata*, and *Madhuca indica*. Of these, focus is on *J. curcas*, which can grow on arid, semi-arid, and wasteland (Figure 1). It requires less water and fertilizer and can survive on infertile soil. It is also pest resistant and has a high yield. Jatropha seed has about 30%–40% of oil content. India has 80–100 million hectares of wasteland, which can be utilized for jatropha cultivation.

The unique properties of bio-diesel, like high cetane, low sulphur content, and better lubricity than petro-diesel, make it an excellent fuel for diesel engines. Tests have indicated the vehicle performance to be satisfactory with blends up to 20% of bio-diesel. There is a marked reduction in carbon monoxide and particulate emissions with the use of bio-diesel. Though blends up to 20% are quite common world over, the Indian manufacturers are apprehensive and recommend only 5% or 10% of bio-diesel in diesel.

Figure 1
Jatropha curcas

IOCL's initiatives

The IOCL (R&D) Centre is the pioneer in the country in undertaking bio-diesel programme in a scientific and systematic manner. Major diesel users were apprised of the programme, and the response from the Indian Railways was particularly encouraging. The Railways sector consumes about 2 MT of diesel, or 5% of the country's production, per year. The Railways sector also has land along the 100 000 km of tracks, which can be utilized for jatropha plantation. So, it agreed to provide 500 hectares of land to the IOCL for planting jatropha. The plantation activity has already started at two sites in Gujarat (Figure 2).

Figure 2
Jatropha plantation by the IOCL on railway land at Surendra Nagar

The IOCL (R&D) Centre has examined and optimized the synthetic processes for producing bio-diesel from various vegetable oils, including rice bran, palm, sunflower, and so on. The process has been patented and scaled up to the pilot level. The technology has been transferred to some firms on non-exclusive basis, on a nominal fee and royalty sharing mechanism. The Centre is also in the process of setting up a continuous bio-diesel pilot plant (Figure 3). Table 1 gives a comparative account of the properties of B5, B20, and diesel.

Figure 3
IOCL R&D bio-diesel pilot plant

Emission studies and field trials

Exhaustive emission testing was carried out at the R&D Centre, on diesel cars and Haryana Roadways buses. Using 10% and 20% blends of bio-diesel in diesel, there was 20% and 25% reduction in carbon monoxide emission and 5% and 10% reduction in particulate matter, respectively, in the cars tested. Similarly, there was a considerable reduction in the smoke emission from buses after using blended diesel.

Tests have been conducted with 5%, 10%, and 20% blends of jatropha bio-diesel in diesel on 16-cylinder Alco Diesel Locomotive Engine for power, specific fuel consumption, firing pressure, and exhaust gas temperature (Figure 4). Trial runs on Shatabdi and Jan Shatabdi express have been carried out with 5% and 10% bio-diesel. A significant reduction in hydrocarbon and carbon monoxide emissions was observed.

Field trials have also been carried out on Tata Indica and LCVs by making them run on blended diesel.

Cost–benefit analysis

The cost of bio-diesel production, as estimated by the Planning Commission of India, is approximately about Rs 16 per litre. It may need subsidy in the form of duty concessions for some time, as is the practice in most countries implementing the bio-diesel programme.

As per the demonstration project of the Planning Commission, in the first phase of the project, jatropha would be planted on 40 000 hectares of land at different places in India.

Table 1 Comparison of B5 and B20 with diesel

Property	*BIS 1460*	*IOCL B5*	*IOCL B20*
Density (g/cm^3)	820–870	834.2	854.7
Carbon residue (% mass)	Max 0.3	0.04	0.01
Sulphated residue (% mass)	Max 0.01	0.001	0.002
Total sulphur (% mass)	Max 0.25	0.02	0.04
Cetane number	Min 48	50.4	54.6
Flash point (°C)	Min 35	39.5	77
Coppor corrosion	Max 1	1	1
Viscosity, 40 °C (cSt)	2.0–5.0	2.497	3.25
Neutralization value (mg/g)	Max 0.2	0.01	0.2
Free glycerin (% mass)	NA	NA	0.001
Total glycerin (% mass)	NA	NA	0.05
CFPP (°C)	Max 6 (W)	0.0	–1
Phosphorus (% mass)	NA	0.001	0.001
Distillation temperature (°C) @ 360 °C	IBP, FBP	96%	96%
Water content (mg/kg)	Max 500	280	240

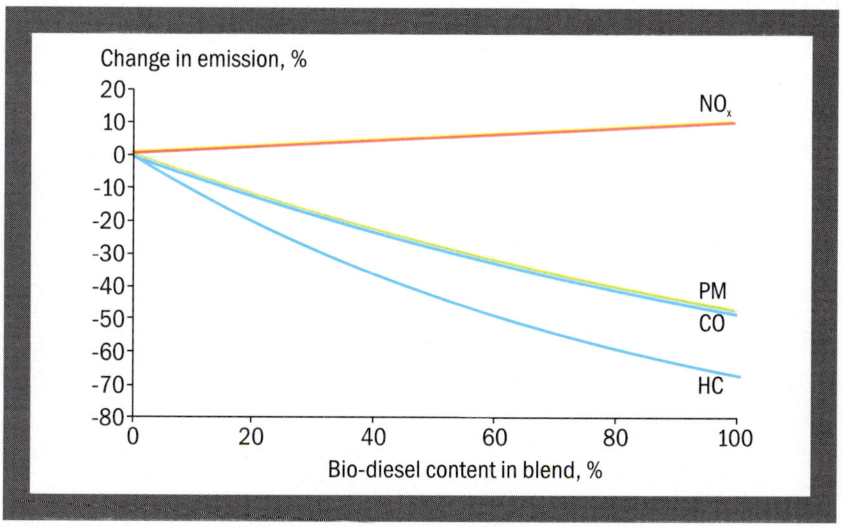

Figure 4 Impact of bio-diesel on emissions from heavy duty engines

Considering per hectare per man (with a family of four people) for the cultivation and seed collection activities, the project could provide employment to 400 000 people in the rural area throughout the year. A yield of 4 MT of seed per hectare at the rate of Rs 5/kg could provide Rs 20 000 to the family in the rural area.

As per the estimate of the Planning Commission, for 5% blending of bio-diesel, approximately 2.62 MTPA bio-diesel would be required. This would require plantation on 2.2 million hectares of land, creating job and livelihood for more than 2 million families.

Benefits

- Bio-diesel is a green, environment-friendly fuel.
- Blending bio-diesel with diesel effectively curbs particulate, smoke, and carbon monoxide emissions.
- Blending also reduces the sulphur level in the fuel and enhances the fuel quality in terms of low aromatics and high lubricity.
- Bio-diesel is produced from non-edible oil plants and creates livelihood options for rural masses.

CASE STUDY AT A GLANCE

- The Indian Oil Company Ltd, established in 1959, merged with the Indian Refineries Ltd in 1964 to form the Indian Oil Corporation Ltd, or IOCL.
- The major products of the company include LPG (liquefied petroleum gas), motor spirit, naphtha, aviation turbine fuel, kerosene, diesel, bitumen, and so on.
- IndianOil's gross turnover for 2007/08 reached a new high of Rs 2 474.79 billion, up by 12.1% from the previous year.
- The IOCL (R&D) Centre is the pioneer in the country in undertaking bio-diesel programme in a scientific and systematic manner.
- The IOCL (R&D) Centre has examined and optimized the synthetic processes for producing bio-diesel from various vegetable oils, including rice bran, palm, sunflower, and so on. The process has been patented and scaled up to the pilot level.
- The Centre is also in the process of setting up a continuous bio-diesel pilot plant.
- Exhaustive emission testing was carried out at the R&D Centre, on diesel cars, Haryana Roadways buses, 16-cylinder Alco Diesel Locomotive Engine, and Tata Indica and LCVs.
- Blending bio-diesel with diesel effectively curbs particulate, smoke, and carbon monoxide emissions.
- Blending also reduces the sulphur level in the fuel and enhances the fuel quality in terms of low aromatics and high lubricity.
- Bio-diesel is produced from non-edible oil plants and creates livelihood options for rural masses.

CASE STUDY 35

JSW Steel Ltd

JSW Steel Ltd

About the company

JSW Group is one of the fastest growing business conglomerates with a strong presence in the core economic sector (Figure 1). This Sajjan Jindal led enterprise has grown from a steel rolling mill in 1982 to a multi-business conglomerate worth $3.7 billion within a short span of time.

As part of the $8 billion O P Jindal Group, JSW Group has diversified interests in steel, energy, minerals and mining, aluminum, infrastructure and logistic, cement and information technology.

India's second largest steelmaker, JSW Steel Ltd, located in Karnataka, consists of the most modern, eco-friendly steel plants with the latest technologies for both upstream and downstream processes.

Environmental profile

The Group is also conscious about its responsibility towards the environment. Therefore, environmental concerns are taken into account during all stages of production and growth.

The JSW Steel Ltd has an integrated policy for addressing the quality, environmental, and safety issues. It has received ISO: 9001 for Quality Management System, ISO: 14001 for EMS (Environment Management System), and OHSAS: 18001 for Occupational Health and Safety Management System.

The company has a well established environmental management department, which is responsible for liaisoning with statutory authorities, coordinating the activities of the EMS, monitoring of environmental quality, managing solid waste, implementing new environment improvement related projects, identifying CDM (Clean Development Mechanism) projects, and so on.

The EMS is responsible for the initiatives taken by the company to reduce environmental impacts of its operations. The effective implementation of the EMS has resulted in improved waste management practices, reduction in water consumption, reduction in overall emissions, and conservation of energy.

Air pollution

Air pollutants emitted from the facility are characterized by SPM (suspended particulate matter), RSPM (respirable suspended particulate matter), SO_2 (sulphur dioxide), and NO_x (oxides of nitrogen). Besides, fugitive emissions, odour, and noise are also generated from the facility.

Air pollution control measures

SPM

The integrated steel plant handles a large volume of solid raw material, which inadvertently emits a lot of dust and particulates. Air pollution control equipment like bag filters, electrostatic precipitators, scrubbers (Figure 2), and dust suppression systems have been installed to control dust pollution. To reduce the emissions of fugitive dust, measures like vacuum sweeping of roads, dust suppression in stock piles, transporting dust in closed tankers, and pneumatic conveying of dust have been introduced. JSW Steel has 100% cast house fume extraction in all its hot metal manufacturing units.

Figure 2
Set of scrubbers and ID fans to clean and pump the gas to the gas holder

RSPM

RSPM is the predominant constituent of process emissions like combustion, fume extraction, and gas cleaning. To abate its emission, gas cleaning is carried out in high pressure drop venture systems; fume extraction systems have been provided with efficient bag filters; and only dust-free fuel gases are used for combustion.

SO_2

The main source of SO_2 is the sulphur present in the raw material and fuel gases. The fuel gases used in the facility, like blast furnace gas, LD gas, and Corex gas, do not contain sulphur. So, SO_2 emission from the JSW Steel Plant is only of the order of 0.63 kg/tcs as compared to more than 1.4 kg/tcs in other similar steel plants.

NO_x

Combustion furnaces are the main source of NO_x emissions. But these emissions are very low, as the fuel gases used do not contain nitrogen.

Odour

The volatile matter in the coke oven plant is the source of odour. The facility uses an environment-friendly technology called non-recovery coking, in which the volatile matter is combusted in the coke oven itself, thus eliminating the problem of odour.

Water pollution and its control measures

The facility has 'near zero' effluent discharge. Treated water is recycled in the plant processes and for other applications like dust suppression. The facility is in the process of installing a reverse osmosis unit for the recovery of water from the waste water.

Hazardous waste

Waste oil is the only hazardous waste generated by the facility, which is sold to authorized recyclers.

Other industrial waste

The main industrial waste generated from the facility comprises Corex sludge, basic oxygen furnace slag and sludge, mill scale, lime fines, coal dust, oxide dust, and lime dust. Corex sludge, basic oxygen furnace sludge, oxide dust, lime dust, and mill scale are reused in the pelletization process. Corex grade slag is used in making cement. Coal dust is reused in the power plant for power generation.

Case study

Reduction of specific energy consumption in the basic oxygen furnace by utilizing off-waste gases from basic oxygen furnaces

Background

Manufacturing steel in integrated steel plants is an energy-intensive process, with each tonne of steel produced consuming 5.5–6.0 Gcal of energy. The energy used accounts for 35%–40% of the cost of production of steel. Therefore, it is necessary that energy conservation measures are taken to make the process of steel manufacture energy efficient and cost-effective.

In the initial stage, Corex units were set up in the facility, where Corex gas with a calorific value of 1800 kcal/Nm^3 was used. Subsequently, blast furnaces were introduced in the facility, where blast furnace gas with a calorific value of 750–800 kcal/Nm^3 was used. This gas couldn't be used alone because of its low calorific value. So, Corex gas was used as a back-up or support fuel. But with the large-scale expansion in steel processing units, there was an overall shortfall of Corex gas.

During the conversion of hot metal to steel in basic oxygen furnace converters, waste gases are emitted. If the converters are operated under suppressed combustion mode (less excess air), CO (carbon monoxide)-rich off-gases are released. These gases were being flared in the facility.

It was realized that these waste gases could be recovered and used as fuel in the furnaces. But there were certain intricacies involved in their usage. Since basic oxygen furnaces operate on batch process, the release of the gases was cyclic. Also, the volume of generation was lower. Further, the CO content in the gases also varied from 10% to 60% because of which it was not possible to use them in a continuous manner. Another problem was related to the dust content. The dust content of the waste gases was of the order 95–110 mg/Nm^3, and this entire dust was let into the atmosphere when the gases were flared.

Process detail

The facility decided to recover the off-gases containing CO in such a way that the calorific value of the recovered gases matched the value of the Corex gas, so that the fluctuation in the calorific value did not affect the power generation and other heating applications. Further, it was decided to supply these gases through the Corex gas grid so as to utilize the existing gas grid and use the blast oxygen furnace gases as a supplement to the Corex gas in all the furnaces. It was envisaged that this would increase the flexibility of the Corex gas and the use of blast furnace gas. Also, the gas cleaning plant treating the waste gases was thoroughly revamped, after which the dust content was reduced to <30 mg/Nm^3, which did not choke the nozzles and burners.

Recovering the off-gases

The off-gases from the converters are extracted at a rate of 110 000 Nm^3/h for 18–20 minutes, though the overall recovery of the usable blast oxygen furnace gases is less than 15 000 Nm^3/h for each converter. It was also necessary to supply these gases at a pressure slightly higher than the Corex gas. Therefore, a gas holder of 30 000 m^3 volume was installed, which accommodated the recovered gases during the generation of off-gases and supplied these gases to the grid during the non-generation period (Figure 3). These gases were fed to the grid through booster pumps at a constant pressure. Since a large part of these gases comprises toxic CO, several interlocks were provided to ensure the safe operation of the system. The off-gases generated in the beginning, containing low amount of CO, and at the end of the cycle were flared by using special flares. To ensure complete flaring, LPG (liquefied petroleum gas)/Corex burners are used.

Figure 3
Gas holder for maximizing gas utilization and reducing gas flow fluctuations

Optimization of the recovery

The calorific value of the gases varies from a very low of 800 kcal/Nm^3 in the initial stage to 2200 kcal/Nm^3 during the middle of the cycle and again drops to <800 kcal/Nm^3 at the end of the cycle. Thus, if the recovery of the off-gases is increased, their calorific value will come down, and if the recovery is restricted, the calorific value will increase. Since the calorific value of the recovered gases is required to match that of the Corex gas, several optimization runs were performed by varying the recovery periods so that maximum energy (product of calorific value and recovered gas volume) is achieved.

Supply of recovered gases to the Corex gas grid

The recovered gases are stored in a gas holder, the volume of which increases with the recovery cycle. Gas boosters of suitable size are used to evacuate the gas before the next batch of making steel starts (Figure 4). This entire process ensures that there is zero flare (Figure 5).

Figure 4
Set of booster fans for maximizing gas evacuation

Figure 5
Zero flare

Cost–benefit analysis

Capital cost of waste gas recovery system: Rs 180 million

Average recovery of waste gases: 70 m^3/tls

Considering 90% achievement of production target, at production capacity of 4 MTPA, the total waste gases generated would be: 25.2 million m^3

Calorific value of the waste gases: 1900 kcal/Nm^3

To generate 1 MW of power, waste gases required: 1400 m^3

Total power generated by utilizing waste gases: 180 000 MWh = 180 MU/annum

Average power cost: Rs 3 per unit

Total amount accrued by power generation: Rs 540 million/annum

Operating cost of gas-based power plant: Re 1/unit

Total operating cost: Rs 180 million/annum

Net amount recovered by selling power: Rs 360 million/annum

Payback period: 6 months

Benefits

- *Better usage of carbon monoxide* The CO being flared earlier is being utilized as a support fuel in the power plant.
- *Saving energy* There has been a reduction in the specific energy consumption in steel making, to the tune of 0.24 Gcal/tcs.
- *Energy conservation* The gases recovered are used as fuel in the existing gas/coal-based power plant, resulting in a considerable amount of resource and energy conservation.
- *Resource conservation* The recovered dust in the form of slurry is being used in the pellet plant (the slurry comprises 60% Fe_2O_3), resulting in resource conservation.
- *Generation of power* By using waste gases in the power plant, 20 MW equivalent of power is being generated, which has resulted in the reduction of 180 000 tonnes of CO_2 emissions per year.
- *Savings* There has been a net saving of Rs 360 million per year by using waste gases for power generation.

CASE STUDY AT A GLANCE

- JSW Steel Ltd is India's second largest steelmaker.
- During the conversion of hot metal to steel in basic oxygen furnace converters, CO (carbon monoxide)-rich off-gases are released. These gases were being flared in the facility.
- The facility decided to recover these waste gases and use as fuel in the furnaces.
- The gases recovered are used as fuel in the existing gas/coal-based power plant, resulting in a considerable amount of resource and energy conservation.
- There has been a reduction in the specific energy consumption in steel making, to the tune of 0.24 Gcal/tcs.
- By using waste gases in the power plant, 20 MW equivalent of power is being generated, which has resulted in the reduction of 180 000 tonnes of carbon dioxide emissions per year.
- The recovered dust in the form of slurry is being used in the pellet plant (the slurry comprises 60% Fe_2O_3), resulting in resource conservation.
- There has been a net saving of Rs 360 million per year by using waste gases for power generation.

ANNEXURE 1 About TERI-BCSD India

Enabling responsible business for a sustainable future

The goal of sustainability is now integral to the long-term survival of human society. Environmental and socio-economic problems are increasing in intensity and complexity. There is a growing appreciation for the need for a collaborative approach to develop strategies for industrial sustainability and corporate responsibility.

In India, a vast body of knowledge and experience is lying untapped within industry, which could be used to implement the sustainability agenda.

Initiated by TERI in 2001, the TERI-BCSD (The Energy and Resources Institute-Business Council for Sustainable Development) India has now evolved into a strong industry body, with membership from diverse sectors, including public sector undertakings, multinationals, and private companies from across India. It is the Indian partner of the WBCSD (World Business Council for Sustainable Development), Geneva.

Mission

To provide an independent and credible platform for corporate leaders to address the issues related to sustainable development and to promote leadership in environmental management, social responsibility, and economic performance.

Goal

To use TERI-BCSD India membership and TERI's research capabilities to gain a better understanding of existing and emerging sustainability concerns facing the industry.

Membership

Membership of TERI-BCSD India is open and is solicited from organizations and not individuals. It is open to any cooperation dedicated to fostering environmental excellence and social responsibility within itself and in the industry as a whole.

For details please visit <www.teriin.orglbcsd>

ANNEXURE 2 Contact details of companies

Aban Power Company Limited
Mr V Ravindran, Chief Operating Officer
25, G N Chetty Road, T.Nagar
Chennai–600 017
Tel.: 044-28157209, 28157210, p9791119911
Fax: 044-28157248
E-mail: ravindran@abanpower.com

ACC Limited
Mr Ulhas V Parlikar, Director-AFR Business, AFR department
Thane Complex, LBS Marg
Thane (West)–400 604
Tel.: 022-25823631/684, 25838413, 09821967941
Fax. 022-67988378
E-mail: ulhas.parlikar@acccement.com, palash.saha@acccement.com

Ayurvet Limited
Dr Deepti Rai, Sr. Project Coordinator
6th floor, Sagar Plaza, district Centre
Laxmi Nagar, Vikas Marg, Delhi 110 092
Tel.: 22455992-94
Fax: 22455991
E-mail: drai@ayurvet.com

Bharat Aluminium Company Ltd.
Mr B K Bhatia, General Manager - HSE
P O Balco Township, Korba–495 684
Tel.: 07759-242054, 99930 00750
Fax: 07759-240476
E-mail: bhawnesh.bhatia@vedanta.co.in

Bharat Aluminium Company Limited
Mr J Ramaswamy, Senior Vice President - QA & Technology
P O Balco Nagar, Korba–495 684
Tel.: 07759-252106
Fax: 07759-241053
E-mail: J Ramaswami@vedanta.co.in

Bharat Aluminium Company Ltd.
Dr P K N Raghavan, General Manager - Alumini Refinery & EHS
P O Balco Township, Korba–495 684
Tel.: 07759-242054
Fax: 07759-240476
E-mail: pkn.raghavan@vedanta.co.in

Chennai Petroleum Corporation Ltd.
Mr N K Rajamani, General Manager (CP, Devt. And R & D)
536, Anna Salai, Teynampet, Chennai–18
Tel.: 044-24342629
Fax: 24341753
E-mail: nkrajamani@cpcl.co.in

Gujarat Co-operative Milk Marketing Federation Limited
Mr B R Mallick, Manager (Marketing)
Amul Dairy Road, Anand–388 001
Tel.: 02692-258506 to 508, 221276, 09375032284
Fax: 02692-240208, 240185
E-mail: brmallick@amul.coop

Hero Honda Motors Ltd.
Mr A K Agarwal, Dy. General Manager
37th KM Stone, Delhi–Jaipur Highway
Sector 33, Gurgaon – 122 001
Tel.: 95124-2372123 to 35 extn. 572, 9810155616
Fax: 95124-2373141, 42

Hindustan Aeronautics Limited
Mr G S Gupta, General Manager
Engine Division
P B No. 9310, Old Madras Road, Bangalore–560 093
Tel.: 080-22866069
Fax: 080-22864027
E-mail: gm.engines@hal-india.com

Hindustan Zinc Limited
Mr Akhilesh Joshi, Chief Operational Officer
Yashad Bhawan, Udaipur–313 004
Tel.: 0294-2421420
Fax: 0294-2424505
E-mail: akhileshjoshi@vedanta.co.in

Honeywell Electrical Devices and Systems India Limited
Mr B Satyanarayanan, Sr. Manager - Production & HOS
'Crescendo', 995-B, Second Avenue, Anna Nagar, Chennai–600 040
Tel.: 044-2681 1234
Fax: 044-2681-1235
E-mail: Sathyanarayanan.b@Honeywell.com

Indian Farmers Fertiliser Cooperative Limited
Mr Anil Kumar Singh, Senior General Manager
IFFCO Aonla Unit
P.O. IFFCO Township
Bareilly – 243 403
Tel.: 0581-2404004/2404003
Fax: 0581-2404227, 05823-234888
E-mail: aksingh-aonla@iffco.nic.in

Indian Oil Corporation Limited
Mr Rohit Bhardwaj, Executive Director (Safety, Health and Environment)
Corporate Office, Scope Complex, Core-2
7, Institutional Area, Lodhi Road, New Delhi – 110 003
Tel.: 24362499
Fax: 24361783
E-mail: bhardwajr@iocl.co.in

Ispat Industries Limited
Mr B K Singh, Executive Director (In change)
Geetapuram, Dolvi, Tal - Pen, Raigad – 402 107
Tel.: 02143-277501 - 277515
Fax: 02143-277553, 277533, 277542
E-mail: bk_singh@ispatind.com

JSW Steel Limited, Vijaynagar Works
Mr S M R Prasad, General Manager (Environment & Energy)
P O Vidyanagar, Toranagallu, Bellary – 583 275
Tel.: 08395-264120, 09448286269
Fax: 08395-264120
E-mail: smr.prasad@jsw.in

Larsen & Toubro Limited
Mr D B Raju, Vice President-Corporate Infrastructure and Services
Gate No. 1, Saki Vihar Road, Powai
Mumbai – 400 072
Tel.: 022-67052497
Fax: 022-67051001
E-mail: rajudb@pgm.ltindia.com

Mangalore Refinery and Petrochemicals Ltd (a Subsidiary of ONGC Ltd)
Mr P P Shenoy, GM (Environment and Safety)
P O Kuthethur, Mangalore – 575 030
Tel.: 0824-2270400 extn.2970
Fax: 0824-2270013
E-mail: shenoypp@mrplindia.com

Nava Bharat Ventures Limited
Y Poornachandra Rao, Executive Vice President
KSIIL Campus (Post), Paloncha, Khammam Dist – 507 154
Tel.: 08744-256015-18
Fax: 08744-256419, 256077
E-mail: nbvlpal@sancharnet.in

NHPC Limited (A Govt. of India Enterprise)
Director (Personnel)
Sector-33, Faridabad 121 003
Tel.: 95129-2278015/2277855
Fax: 95129-2258211

Northern Coalfields Limited
Mr Niranjan Das, Director (Tech./P & P)
P O Singrauli Colliery, Dist. Singrauli – 486 889
Tel.: 07805-266 621, 07805-256201
Fax: 07805-266456
E-mail: dtpp.ncl@nclhq.nic.in

Numaligargh Refinery Limited
Mr R Baruah, Dy. General Manager 9TS & Inspection
Numaligargh, Golaghat – 785 699
Tel.: 03776-265416
Fax: 03776-265578
E-mail: manas.r.barua@nrl.co.in

Numaligarh Refinery Limited
Mr Ray Baruah, Dy. General Manager (TS & Inspection)
Numaligarh, Dist. Golaghat – 785 699
Tel.: 03776-265416, 09435054232
Fax: 03776-265578
E-mail: manas.r.barua@nrl.co.in

Orchid Chemicals and Pharmaceuticals Limited
Mr V S Padalkar, Vice President-Projects & Engineering
Plot No.L8 (Part) & L9, MIDC Area,
Waluj, Aurangabad – 431 136
Tel.: 0240-2554992/3/4/5/, 2567399, 2567111
Fax: 0240-2554968
E-mail: padlkar@orchidpharma.com

Orissa Power Generation Corporation Ltd.
Mr S E Sekar Director (Operation)
Ib Thermal Power Station, Baharpali, Jharsuguda–768 234
Tel.: 06645 222220, 9937260013
Fax: 06645 222230
E-mail: sekar.sk@opgc.co.in

Rashtriya Ispat Nigam Limited, Vishakhapatnam Steel Plant
Mr V S Rao, General Manager (Environment & Safety)
Visakhapatnam–530 031
Tel.: 0891-2518301
Fax: 0891-2518237
E-mail: cmd@vizagsteel.com, pbcrao@vizagsteel.com

Reliance Industries Limited
Mr Sanjay J Patel, HOD - Environement Dept.
P O Petrochemicals
Vadodara – 391 346
Tel.: 0265-6696467, 09998213412
Fax: 0265-6693895
E-mail: sanjay.j.patel@ril.com

Sesa Goa Limited
Mr Mahesh Patil, General Manager (Environment)
Sesa Ghor, 20 EDC Complex, Patto, Panaji – 403 001
Tel.: 0832-2460710
Fax: 0832-2460719
E-mail: mpatil@sesagoa.com

Shree Cement Limited
Mr P K Tripathy, Sr. Vice President
Bangur Nagar, Beawar – 305 901
Tel.: 01462-228101-106
Fax: 01462-228117-119
E-mail: tripathypk@shreecementltd.com

Sterlite Industries (India) Limited
Mr Ramesh Nair, Chief Operating Officer
Madurai Bypass Road, T V Puram P O
Tuticorin – 628 002
Tel.: 0461-6612591, 0461-6612143
Fax: 0461-2340129
E-mail: ramesh.nair@vedanta.co.in

Sudarshan Chemical Industries Limited
Mr K R Ambekar, V P (Engg. & EHS)
46, MIDC Estate, Dhatav, Roha, Raigad – 402 116
Tel.: 02194-263531
Fax: 02194-263602
E-mail: krambekar@sudarshan.com

Tamil Nadu Newsprint and Papers Limited (A Government of Tamil Nadu Enterprise)
Mr R Mani, General Manager (Operations)
Kagithapuram, Karur District – 639 136
Tel.: 04324-27701-277010, 09442591315
Fax: 04324-277027 & 277285
E-mail: mani.r@tnpl.co.in

Tata Steel Limited
Mr Shashi Bhushan Prasad, Head-Environment & OHS
Tata Steel Limited
Steel Works, Jamshedpur–831 001
Tel.: 0657-2424125
Fax: 0657-2427819
E-mail: sb.prasad@tatasteel.com

The Singareni Collieries Company Limited
Mr S Narsing Rao, Chairman and Managing Director
Singareni Bhawan, P B No. 18, Red Hills
Khairatabad, Hyderabad – 500 004
Tel.: 040-23393746
Fax: 040-23393746
E-mail: cmd@scclmines.com

United Phosphorus Limited
Mr Raju Shroff, Chairman
Uniphos House
C D Marg, 11th Road, Madhu Park, Khar (W)
Mumbai 400 052
Tel.: 9820082222, 022-26468311/322
Fax: 022-26041010
E-mail: shroffrd@uniphos.com

Vikram Cement (A Unit of Grasim Industries Ltd)
Mr R M Gupta Sr. Executive President
P O Khor, Dist. Neemuch – 458 470
Tel.: 07420-235568
Fax: 07420-235524
E-mail: radhamohan.gupta@adityabirla.com

Wipro Limited
Mr Chitra P B Gowda, EHS Manager
#70/1,2,3,4 Keonics Electronics city, Doddathoguru Village
Bangalore–560 100
Tel.: 080-30293864
Fax: 080-30293900
E-mail: Chitra.byregowda@wipro.com

Wipro Technologies
Mr M P Nagendra, EHS Manager
No. 88, S B Towers, M G Road, Bangalore–580 001
Tel.: 080-41366008/41366086
Fax: 080-25586657
E-mail: nagendra.mysore@wipro.com